Unhomely Life

RGS-IBG Book Series

For further information about the series and a full list of published and forthcoming titles please visit www.rgsbookseries.com

Published

How Cities Learn: Tracing Bus Rapid Transit in South Africa
Astrid Wood

Defensible Space on the Move: Mobilisation in English Housing Policy and Practice
Loretta Lees and Elanor Warwick

Geomorphology and the Carbon Cycle
Martin Evans

The Unsettling Outdoors: Environmental Estrangement in Everyday Life
Russell Hitchings

Respatialising Finance: Power, Politics and Offshore Renminbi Market Making in London
Sarah Hall

Bodies, Affects, Politics: The Clash of Bodily Regimes
Steve Pile

Home SOS: Gender, Violence, and Survival in Crisis Ordinary Cambodia
Katherine Brickell

Geographies of Anticolonialism: Political Networks Across and Beyond South India, c. 1900–1930
Andrew Davies

Geopolitics and the Event: Rethinking Britain's Iraq War through Art
Alan Ingram

On Shifting Foundations: State Rescaling, Policy Experimentation And Economic Restructuring In Post-1949 China
Kean Fan Lim

Global Asian City: Migration, Desire and the Politics of Encounter in 21st Century Seoul
Francis L. Collins

Transnational Geographies Of The Heart: Intimate Subjectivities In A Globalizing City
Katie Walsh

Cryptic Concrete: A Subterranean Journey Into Cold War Germany
Ian Klinke

Work-Life Advantage: Sustaining Regional Learning and Innovation
Al James

Pathological Lives: Disease, Space and Biopolitics
Steve Hinchliffe, Nick Bingham, John Allen and Simon Carter

Smoking Geographies: Space, Place and Tobacco
Ross Barnett, Graham Moon, Jamie Pearce, Lee Thompson and Liz Twigg

Rehearsing the State: The Political Practices of the Tibetan Government-in-Exile
Fiona McConnell

Nothing Personal? Geographies of Governing and Activism in the British Asylum System
Nick Gill

Articulations of Capital: Global Production Networks and Regional Transformations
John Pickles and Adrian Smith, with Robert Begg, Milan Buček, Poli Roukova and Rudolf Pástor

Metropolitan Preoccupations: The Spatial Politics of Squatting in Berlin
Alexander Vasudevan

Everyday Peace? Politics, Citizenship and Muslim Lives in India
Philippa Williams

Assembling Export Markets: The Making and Unmaking of Global Food Connections in West Africa
Stefan Ouma

Africa's Information Revolution: Technical Regimes and Production Networks in South Africa and Tanzania
James T. Murphy and Pádraig Carmody

Origination: The Geographies of Brands and Branding
Andy Pike

In the Nature of Landscape: Cultural Geography on the Norfolk Broads
David Matless

Geopolitics and Expertise: Knowledge and Authority in European Diplomacy
Merje Kuus

Everyday Moral Economies: Food, Politics and Scale in Cuba
Marisa Wilson

Material Politics: Disputes Along the Pipeline
Andrew Barry

Fashioning Globalisation: New Zealand Design, Working Women and the Cultural Economy
Maureen Molloy and Wendy Larner

Working Lives - Gender, Migration and Employment in Britain, 1945-2007
Linda McDowell

Dunes: Dynamics, Morphology and Geological History
Andrew Warren

Spatial Politics: Essays for Doreen Massey
Edited by David Featherstone and Joe Painter

The Improvised State: Sovereignty, Performance and Agency in Dayton Bosnia
Alex Jeffrey

Learning the City: Knowledge and Translocal Assemblage
Colin McFarlane

Globalizing Responsibility: The Political Rationalities of Ethical Consumption
Clive Barnett, Paul Cloke, Nick Clarke & Alice Malpass

Domesticating Neo-Liberalism: Spaces of Economic Practice and Social Reproduction in Post-Socialist Cities
Alison Stenning, Adrian Smith, Alena Rochovská and Dariusz Świątek

Swept Up Lives? Re-envisioning the Homeless City
Paul Cloke, Jon May and Sarah Johnsen

Aerial Life: Spaces, Mobilities, Affects
Peter Adey

Millionaire Migrants: Trans-Pacific Life Lines
David Ley

State, Science and the Skies: Governmentalities of the British Atmosphere
Mark Whitehead

Complex Locations: Women's geographical work in the UK 1850–1970
Avril Maddrell

Value Chain Struggles: Institutions and Governance in the Plantation Districts of South India
Jeff Neilson and Bill Pritchard

Queer Visibilities: Space, Identity and Interaction in Cape Town
Andrew Tucker

Arsenic Pollution: A Global Synthesis
Peter Ravenscroft, Hugh Brammer and Keith Richards

Resistance, Space and Political Identities: The Making of Counter-Global Networks
David Featherstone

Mental Health and Social Space: Towards Inclusionary Geographies?
Hester Parr

Climate and Society in Colonial Mexico: A Study in Vulnerability
Georgina H. Endfield

Geochemical Sediments and Landscapes
Edited by David J. Nash and Sue J. McLaren

Driving Spaces: A Cultural-Historical Geography of England's M1 Motorway
Peter Merriman

Badlands of the Republic: Space, Politics and Urban Policy
Mustafa Dikeç

Geomorphology of Upland Peat: Erosion, Form and Landscape Change
Martin Evans and Jeff Warburton

Spaces of Colonialism: Delhi's Urban Governmentalities
Stephen Legg

People/States/Territories
Rhys Jones

Publics and the City
Kurt Iveson

After the Three Italies: Wealth, Inequality and Industrial Change
Mick Dunford and Lidia Greco

Putting Workfare in Place
Peter Sunley, Ron Martin and Corinne Nativel

Domicile and Diaspora
Alison Blunt

Geographies and Moralities
Edited by Roger Lee and David M. Smith

Military Geographies
Rachel Woodward

A New Deal for Transport?
Edited by Iain Docherty and Jon Shaw

Geographies of British Modernity
Edited by David Gilbert, David Matless and Brian Short

Lost Geographies of Power
John Allen

Globalizing South China
Carolyn L. Cartier

Geomorphological Processes and Landscape Change: Britain in the Last 1000 Years
Edited by David L. Higgitt and E. Mark Lee

Unhomely Life

Modernity, Mobilities, and the Making of Home in China

Xiaobo Su
Department of Geography
University of Oregon

WILEY

This edition first published 2024

Registered Offices
John Wiley & Sons, Inc., 111 River Street, Hoboken, NJ 07030, USA
John Wiley & Sons Ltd, The Atrium, Southern Gate, Chichester, West Sussex, PO19 8SQ, UK

For details of our global editorial offices, customer services, and more information about Wiley products visit us at www.wiley.com.

Wiley also publishes its books in a variety of electronic formats and by print-on-demand. Some content that appears in standard print versions of this book may not be available in other formats.

Library of Congress Cataloging-in-Publication Data is applied for:
Paperback ISBN: 9781394176304
Hardback ISBN: 9781394176298

Cover Design: Wiley
Cover Image: © Xiaobo Su

Set in 10/12pt Plantin Std by Straive, Pondicherry, India
Printed and bound by CPI Group (UK) Ltd, Croydon, CR0 4YY

C9781394176298_210324

To Huang Chunyuan and Su Yufei

Contents

Preface and Acknowledgments

It is said that we geographers write about other people's stories, which somehow reflect our own consciousness. It is also said that literally, geography is about earth writing, or writing about the conjunction between the physical existence of nature and the human practices of everyday life. Located at the center of this conjunction, home is imbued with stories.

In this book, I narrate various stories of homemaking and unmaking in the context of China's great transformation in the past four decades. My interest here is not so much in analyzing China's housing market disoriented from the use value of dwelling to the exchange value of economic speculation, nor in highlighting the ideology of house ownership as a symbol of middle-class society, as if such gestures are self-evident in the race for wealth and fame. Rather, I am more concerned with the ongoing deal between profit and meaning that currently affects those who strive for homemaking as an integral part of their daily life in China. The question of "what is home" cannot be separated from the question of "where is a place to call home." Inserted directly into the actual and imaginative space of daily life, homemaking goes beyond the material forms of private residence to reach deep into interior subjectivity and social consciousness that become increasingly difficult to reconcile with the tenets of rational efficiency prevalent in an era of speedy modernization.

My long interest in studying home, and especially its multiple meanings to people on the move, must in one way or another result from my own experiences of migration and travel. My birth-family house, located in a rural village in central China, has been permanently left behind in my forward journey, ever since my parents moved from it to become typical rural-to-urban migrants when I was 14 years old. Although the birth house lingers in my dreams, I dare not visit it in person. To some people, the house of one's birth in relation to childhood and old buddies entails symbolic power, either as a place of origin or a sentimental embodiment of a time gone by. To me, it is a starting point from which I embark upon a never-returning journey. By abandoning my birth house and childhood years, I have lost my rooted home. Later, I drifted yet, farther away, until I became a transnational migrant in the United States. Possibly I can define myself as a subject of liquid modernity.

Shuttling back and forth between the United States and China, I somehow lose a rich sense of feeling at home in either geography or society. Rather, I develop an unruly sense of distance from wherever I stay, and look for new possibilities there. In Eugene, Oregon, I leave my house for work in the morning and come home at the end of the day. During holidays, I become a tourist to travel around in the United States and then go home at the end of the journey. I must confess that I have not been bothered by any housing problem. Nevertheless, a typical US middle-class lifestyle cannot guarantee me a sense of home. The most recent upheaval of Asian hatred after 2016 causes me much mental stress and fearfully reminds me of racial inequality and discrimination. In the United States, a country where I have regularly dwelt, I define myself as an alien of color with permanent residency, indicating that I choose to be exiled to a land I seem to be familiar with.

Almost every year before the COVID-19 pandemic, I traveled back to China in summers. This was and remains a passage of "going home" (huijia, 回家), but the country's dramatic change always makes me feel both excited and alienated. In China, I develop a mood of being a guest, either because everyone treats me as a foreigner from the United States, or because I stay in one hotel or another. Despite my close ties and frequent travels, I have increasingly felt myself to be an alien in the country of my birth. To me, one form of going home means going to the place where my parents reside. They are my biological root and cultural mooring in this highly mobile world of which I am an essential part. The versatile homes that they make in their life enable me to develop a sense of belonging and comfort, no matter how far I have moved away from them. Ironically, I cannot spend much time accompanying them during my stay in China, since I am kept busy with fieldwork on homemaking by others or visiting friends for enjoyment. To my parents, my arrival at home means a forthcoming departure in a week or two. From their perspective, I am their most familiar stranger. Once strangeness starts to shape my familiarity with China, I slide into an unhomely condition, which means struggling between a loss of home and the search for home.

The loss of cultural roots leads me to think that there is an urgent need to understand the lack of feeling at home as a common phenomenon. Indeed, everybody has become unhomely in one way or another because of globalization, increasing mobility, forced displacement, or other causes. Here, unhomeliness is not simply ontologized as a consequence of structural transformation, but also remains a universal experience that we share in common, in both psychological and cultural ways. Ever since the Industrial Revolution, capitalist modernity has removed the very forms of rootedness in time and space and thus rendered impossible the establishment of emotional attachment to a stable society or an inclusive community. Or as Karl Marx says, all that is solid melts into air for the sake of capital accumulation. Mobilities are decisive in my own unsettled sense of home in adult life and render my journey unhomely. Unsurprisingly this sense inflects the conceptual thinking and empirical analysis that follow in this book.

My account goes beyond a development approach that would define the condition of being unhomely as the consequence of economic modernization or the pain of China's dramatic transformation into a global economic power. Even the ills of modernity, as many scholars frequently cite, cannot offer a nuanced description of how individuals react to China's great transformation. Instead, I use the notion of unhomely life as a unique lens to examine more closely the subjective experiences of those on the move in order to understand why they experience the loss of home feelings and how they search for home through imaginations and practices to counter such a loss. Therefore, my analysis shifts from the domestic space of household responsibility and the materiality of housing and property ownership to much broader experiences of homelessness derived from time-space discipline in an era of speedy modernization. In one specific place, Lijiang, a well-known tourist destination in Yunnan province, China, I have witnessed for two decades how individuals with different social and economic backgrounds (gender, age, ethnicity, and class) oscillate between meaning and profit in their everyday life to make and unmake their home.

Unhomely Life provides a critical analysis of how mobile individuals in China make home for a better life in a rapidly changing world. These mobile individuals, including lifestyle migrants and retreat tourists from China's big cities, displaced native town residents, and local service workers from rural Lijiang, engage in an endless journey to mitigate the gap between the home they leave behind or escape from, the existential home in which they currently dwell, and the ideal home that they imagine for retreat and comfort. Filled with compromise, illusion, and resistance, their journey epitomizes an unhomely life through which they make efforts to negotiate for a relatively comfortable existence under a given set of constraining circumstances in China. I argue that the making of home is not confined to a peculiarly economic, rational calculation for maximum return, but expresses a refusal of the *status quo* and represents resistance against and compromise with the disappointing conditions of modernity. This synthesis of resistance and compromise in the process of homemaking has profound implications for our understanding of the trajectories of social transformation in China.

Theorizing back to the critical geographies of home, I emphasize that the loss of home that ordinary Chinese people experience has already occurred to Europeans following the Industrial Revolution, and to individuals in colonial and postcolonial societies, and continues to haunt them today. Thus, the loss of home, or a sense of homelessness, is not exceptional to China, but becomes a universal reality derived from capitalist modernity. What matters is how the loss of home spatially and temporally unfolds in various geographical contexts, and how individuals mitigate such a loss. As one of the most idealized sites of human existence, home offers a critical entry point into understanding and unraveling the micro-geographies of social and spatial uncertainty in a rapidly changing world. I deploy the vocabulary of the loss of home feelings and longing for home throughout the book to illuminate an in-between condition that reflects and reinforces the unevenness of mobilities and modernity. As far as home is concerned, we must critically analyze the practices of everyday life through a dialectic relation between here and there, between past and present, rather than in static isolation.

Over the past two decades, I have accumulated enormous debts of gratitude to many scholars and friends who have helped me to furnish and articulate the in-between condition between home and homelessness. I started my Lijiang adventure in 2002 when I worked with Jigang Bao. Being an enthusiastic mentor, Bao always supports my theoretical exploration and pushes me to ask critical research questions. In Singapore, my former advisor Peggy Teo inspired me to be a critical geographer so that I can expand beyond political discourses in China to think about social and economic inequality that underpins the practices of everyday life. Likewise, Henry Yeung always urges me to go beyond the case study of China and theorize back, in this case, a critical understanding of home.

In the process of writing this book, I have received valuable support and comments from Kai Bai, Weijun Chen, Elaine Ho, Lily Kong, Kean Fan Lim, Jianxiong Ma, Laurence Ma, and Shanyi Zhou. Critical and detailed comments from Alec Murphy and two anonymous reviewers pushed me to build a dialogue with the critical geographies of home and clarify the uniqueness of a Chinese conception of home. In May and June 2023, I had two conversations with Daniel Sui, who expressed great interest in this book and highlighted to me the significance of big pictures and theorization in academic research. My two research assistants, Qingyan Huang and Jiayu Yuan, helped me enormously when I was unable to return to Lijiang to do fieldwork during the COVID-19 pandemic. Friends and respondents in Lijiang generously shared with me their life stories interesting enough to sustain my interest in the town for two decades. Without their generosity, I could never have completed this book. Gratitude is also due to my colleagues in the Department of Geography, University of Oregon, who shared many administrative burdens at a key point in the completion of this manuscript. I benefit tremendously from stimulating conversations with Amy Lobben, Andrew Marcus, and Karen Ford. I am particularly indebted to Chih Yuan Woon and Ruth Craggs, the two editors of this book series, for their perceptive and thoughtful comments on the final manuscript. I feel grateful to Mary Sharon Moore who offers superb proofreading and Rebecca Walter and Carol Thomas who assisted in preparing the manuscript for production. Portions of chapter 3 are revised from my article "Lifestyle migration and the (un)making of ideal home," *Geoforum*, 2020, 115: 111–119, reproduced with permission from Elsevier.

Writing a book on homemaking cannot be possible without the support of my own family members who foster a homely environment for me. Despite my feeling stuck in an in-between condition encapsulating the United States and China, my wife Huang Chunyuan and son Su Yufei create an oasis of calm that enables me to slow down to write about homemaking. I would not have written this book without their affective company, particularly during the COVID-19 pandemic. My mother, Tang Xiangzhen, and father, Su Yeshan, anchor my liquid life journey around them so that I do not completely get lost. My mother-in-law, Zhong Meizhen, and father-in-law, Huang Bingqian, demonstrate to me how cyclical time currently plays out in China's *xiangtu* or rural society, and provide me with the best hospitality I have ever enjoyed in my life. Although my parents and in-laws have no idea of what I am writing about, they always believe that I can do something good to benefit society. For decades, my brother Su Boqing, and two sisters, Su Yan'e and Chen Yanli, have given me unfettered warmth and comfort, which sustain my sense of belonging. It is to my family members that I dedicate this book.

Notes on Fieldwork

My research on Lijiang, Yunnan Province in China, started in 2002 and spanned almost two decades. The bulk of data collection in Lijiang was conducted during nine intervals between 2002 to 2020. In 2004 and 2005, I spent ten months doing intensive ethnographic fieldwork in Lijiang, including over 150 in-depth interviews with tourists, migrants, service workers, and native residents. Then I made five return visits (2007, 2012, 2015, 2016, 2018) to Lijiang and each visit lasted between one and five weeks. The most recent fieldwork in October 2020 was conducted by my research assistant Huang Qingyan, because the COVID-19 pandemic stopped me from returning to China with ease. This book largely relies on the data collected since 2015 to exemplify the town's most recent transformation. Nevertheless, I relied on the earlier data I collected in the 2000s when I focused on native residents (particularly in Chapter 5), because almost all of them in the town had moved to the new city and rented out their property to domestic migrants.

Since 2015, I have interviewed forty-five lifestyle migrants (twenty-five men and twenty women), forty domestic tourists (twenty-two men and eighteen women), twenty service workers (eight men and twelve women), and five native residents. For lifestyle migrants, they had lived in Lijiang for at least one year prior to the interview. Half of them had spent more than five years in Lijiang. By definition, they are migrants rather than tourists, as they leave the cities where they used to live and stay for over six months in Lijiang to establish tourism-related businesses such as cafés, guesthouses, outdoors shops, and bars. Most of them were born in the 1970s and 1980s, and a few in the 1990s. All of them are Han ethnicity. In terms of their socioeconomic position within Chinese society, their previous employment was predominantly in either the public sector as teachers or civil servants, or in the business sector as professionals or managers. A few young migrants originated from relatively affluent families, allowing them to receive financial support from their parents to operate businesses in the town. Lifestyle migrants rarely relocate their household registration (or *hukou* in Chinese) to Lijiang, but their school-age children can go to local public primary schools since the local government wants to retain these migrants. Having said this, many choose to send their children back to the cities where they used to live for the purpose of a better education in middle and high schools, and some others simply leave Lijiang once their children are old enough to go to school.

Tourist respondents came from China's big cities and generally spent at least one week in Lijiang Old Town. Their age ranged from late twenties to early fifties. Middle-class independent tourists (in contrast to group tourists, business tourists, or young budget tourists) were strategically chosen in order to address how they critically understand the meanings of home when they are away from their regular abode. Local service workers originated from urban villages in Lijiang's new city or rural villages with a manageable distance to the old town. Their age ranged from mid-twenties to fifties, and all were married. Among the twenty service workers I interviewed, 70% are Naxi and 30% Bai, two major ethnic minority groups in Lijiang.

Regarding interviews with lifestyle migrants and service workers, I specifically identified a nonrandom selection of respondents who could offer distinctive perspectives about daily experiences in Lijiang. Then I asked them to introduce me to other respondents so that I could expand the relationship of basic trust to gather further information. Alternatively, I approached tourist respondents randomly in restaurants, guesthouses, and streets. Respondents were not expected to represent a whole sample of domestic tourists in Lijiang, nor did their interviews generate exhaustive knowledge about tourist experiences. All interviews ranged from forty-five to ninety minutes in length, and were conducted in Mandarin. Stories can elicit relevant details with an emphasis on the density of meanings and reflections by respondents. This storytelling format allowed respondents to narrate their experiences in their own way.

In Lijiang, people generally love to talk about their life with unacquainted persons in a relaxed touristic setting where social obligation is relieved. With the exception of three respondents who had not much time to be interviewed because they rushed to deliver and collect parcels in the town, other respondents willingly accepted our interview request and elaborated their life stories. Some even actively approached us for an interview when they noticed that we were collecting life stories. Some others were interviewed more than once.

To understand the broad context of Chinese society, I also relied on my own experience of living in big cities. My regular stay in Guangzhou every summer and frequent travel to Beijing, Shenzhen, and Shanghai before 2020 allow me to develop a nuanced impression of hectic life in these first-tier cities in China, and my interactions with friends in these cities further increasing my understanding of how urban living has been accelerated for efficiency. In other words, these experiences allow me to resonate with my respondents (tourists and lifestyle migrants) in Lijiang regarding speed and busyness, but I did not conduct any participant observations about my respondents' actual life in other cities. Indeed, this observation can generate critical insights into the intimate and uncanny search of paradise, as shown by Li Zhang's (2012) critical analysis of homemaking by middle-class citizens in Kunming, the capital city ofYunnan province.

Doing fieldwork in Lijiang only is a methodological choice, as I prefer to approach my respondents in one place rather than reaching out to them in their regular settings. It would be ideal if I could trace back to the daily life of my respondents in metropolitan cities or rural villages. It can be a difficult undertaking to visit different cities and villages to approach people who willingly share their time with an unfamiliar scholar for meaningful reflections on their personal life. Most tourist respondents have been busy in their regular cities, including work, family affairs, Internet surfing, and socialization, in order to efficiently use every minute. Some respondents explicitly told me that they never wanted to talk about themselves if they did not meet me in Lijiang. Henceforth, I maintained Lijiang as a field of interrelationships in which people from cities and villages converge. I heed recent calls for the territorial and relational conception of place, and understand Lijiang as a contact zone with "dense bundles of social relations and power-infused interactions that are always formed out of entanglements and connections with dynamics at work" (Hart 2002, p. 297). Even though data collection was grounded in Lijiang, the fieldwork took place within a translocal cultural setting related to the social and spatial networks that connect my respondents with other Chinese cities.

Scholars talk about conducting fieldwork in sites of constant contact, such as motels or airports, which are designed to accommodate mobile people (Clifford 1997) or night bars for intense discussion (Appadurai 1991). Lijiang is such a mobile place, as people constantly come and go. This mixture between short-lived and flimsy interactions among people on the move, and the long-lived and stable sediment in Naxi traditional society presents a complicated and contradictory parameter of daily life. Tourists and lifestyle migrants said that they enjoy Lijiang's lifestyle very much and indeed, contributed culturally or commercially to myth making in Lijiang, but very few actually stay long enough to live a settled way of life. To the respondents I interacted with, regardless of their background, Lijiang is like the site of a grown-ups playground, or as Levi-Straus puts it, "a place of transit, not of residence" (cites from Clifford, 1997, p. 17). In such a place, the only unchanged thing is the (un)making of home for meaning or money (Fog Olwig and Hastrup 1997). Meanwhile, native residents and Lijiang locals avoid the old town and leave it to outsiders. To them, the town is a place from which they earn money, where their encounters with others are fleeting and arbitrary. In such a place where people constantly come and go, stay and departure are in rapid flux. Maybe I am one of the very few who retain such a long passion with the town, spanning nearly two decades. During this period, I can see that the old town and Lijiang's new city, like many other Chinese cities, have undergone dramatic changes, but some elements remain rooted. This two-decade-long engagement with Lijiang enables me to observe the continuity and disruption born out of China's dramatic transformation in the post-Mao era.

Chapter One
Introduction: From *Xiangtu* China to Unhomely China

"Lijiang is my home and I enjoy it very much. Do I want to go back to Beijing? No, absolutely no! I cannot endure living in Beijing anymore." In her well-maintained courtyard in Lijiang Old Town, a popular destination located in Yunnan province, China, and known for its relaxed lifestyle and Naxi ethnic culture, Mahua expresses to me her feeling of being at home in Lijiang and her disfavor with Beijing. After working in Beijing for years, Mahua became burned out by the hectic pace of life and high-pressure work. In 2013, she migrated to Lijiang and spent her savings leasing a courtyard house, turning it into a five-room guesthouse. After hiring a local female villager as a domestic worker in the guesthouse, Mahua has plenty of time to enjoy a simple and relaxed lifestyle, and shares it with those tourists who want to temporarily escape the hustle and bustle of daily life in China's big cities such as Beijing and Shanghai. Indeed, she hopes her guests will retreat into a space of slowness and simplicity in a world rife with speed and excitement. As she explains, "I enjoy chatting with my guests and listening to their stories. In big cities, they cannot find a listener, since everyone is too busy to listen to others. By offering this sort of friendship, I rely on returning customers and their friends. They come to stay here again and again. There is something they cannot find in big cities, but can experience in Lijiang."

The thing that Mahua holds dear to her heart is the home, or specifically, a combination of feeling at home and experiencing a homely way of life, even though she does not own the guesthouse, nor has she relocated her household registration to Lijiang. As Mahua explains, it took her nearly two years in Lijiang

Unhomely Life: Modernity, Mobilities, and the Making of Home in China, First Edition. Xiaobo Su.

to recover from mental fatigue and discard the nightmare of rushing to complete job tasks. In Beijing, she felt dreadful and lonely, since she worked so hard that she had no time to slow down or build meaningful relations with the city. Her hometown in Jiangsu province in eastern China has undergone dramatic transformation, becoming an unfamiliar place that she does not want to return to, even though her parents expect her to do so. Between a hectic city hollowed out of meanings and a hometown left behind, Mahua had no place to call home and experienced a loss of home feelings. To counter such a loss, she migrated to Lijiang to regain a sense of home in a tourist destination in rapid flux. Ironically, this sense of home has become a good selling point for attracting millions of domestic tourists who more or less suffer from a similar loss of home feelings, as Mahua had before, in China's big cities.

For two decades I have been intrigued by stories shared by Mahua and other respondents about how they enjoy Lijiang's lifestyle and even treat Lijiang Old Town as an ideal home. Their literal meaning of home points to moments of homelessness, or a loss of "home feelings," that plague them in China's big cities and push them to stay in Lijiang for a sense of home, temporally or perennially. This felt sense of homelessness is not the same as being actually unhoused or lacking decent physical shelter, but highlights a loss of home feelings due to hectic daily schedules, high work pressure, and a disconnection with surroundings. Coping with such a loss in Lijiang by Mahua and others from China's big cities, however, results in the displacement of native residents from their courtyard houses and the supply of service by workers from surrounding villages.[1] The making of one home corresponds to the unmaking of the other. As I accumulated life stories from respondents in Lijiang, I came to comprehend an insidious crisis of home in contemporary China. The faster that Chinese society becomes modernized, the less individuals feel at home, and the more they yearn for a sense of home. This central paradox forms the core of *Unhomely Life.*

This paradox deserves careful attention since it spells out a situation of seesawing between the loss of home and the search for home. I conceptualize this situation as "unhomely life." Unhomely life is not exceptional to Mahua and other middle-class residents in Chinese cities, but becomes a common reality derived from modernity. The most dramatic consequence of modernity is the formation of homeless mind, that is, a feeling of psychological homelessness (Berger, Berger, and Kellner 1974). Thanks to modernizing forces embedded in technocratic rationality, individuals are removed from the cultural root of belonging and familiarity, and become exposed to alienation and strangeness. As Berger et al. (1974, p. 82, original emphasis) state, "*modern man has suffered from a deepening condition of 'homelessness.'* The correlate of the migratory character of his [sic] experience of society and of self has been what might be called a metaphysical loss of 'home.'" Similarly, Berman (1982, p. 35) states that "Our desire to be rooted in a stable and coherent personal and social past, and our insatiable desire for growth—not merely for economic growth but for growth in experience, in pleasure, in knowledge, in sensibility [...] destroys both the physical and social landscapes of our past, and our emotional links with those lost worlds." These scholars focus mainly on the experiences of homelessness in Europe after the Industrial Revolution, yet China is no exception. What matters is how the loss of home feelings spatially and temporally unfold in different geographical contexts, and how individuals imagine and experience a homely life to mitigate such a loss.

As Mahua indicated and I will emphasize, a temporal perspective is critical to understanding the relationship between home and modernity (Adam 1998; Friedman 2020; Heller 1999). To Virilio (1986), modernity is characterized by speed. It is a world in motion to accelerate the pace of machines and people, giving rise to a society trapped on a perpetual race course. Once efficiency becomes a constitutive element of modern capitalist society, the pace of economic production and social reproduction becomes inevitably accelerated to facilitate the logic of capital accumulation. Or as Rosa (2013, p. 161) argues, in the capitalist economic system, acceleration becomes "an inescapable compulsion embedded in the material structures of society." No society is more obsessed with speed than China. If the automobile represents a form of speed on wheels in Western societies, the high-speed train in China elevates the addiction to speed to an unprecedented level.[2] The outcome of speedy modernization is not always rosy as expected, particularly with regard to home.

No one can underestimate the significance of home as a key cultural value in China since it touches upon the deep root of Chinese society. For instance, Confucius's dictum that "while one's parents are alive, one should not travel to distant places" reinforces a strong commitment to homemaking for the purpose of filial piety;[3] the popular Chinese aphorism "fallen leaves return to the roots" (叶落归根, yeluo guigen) reflects that home is an ultimate destination for mobile persons to ground themselves. Home as a kernel in Chinese culture builds upon China's rural foundation. Over 70 years ago, renowned Chinese anthropologist Fei Xiaotong published *Xiangtu Zhongguo* (literally, soil-bound China, 乡土中国) to analyze the structural foundations of Chinese society.[4] At the beginning of his book, Fei (1992, p. 37) specifies that "Chinese society is fundamentally rural. I say that it is fundamentally rural because its foundation is rural." The closeness to soil as well as national environment characterizes this rurality. Related to this rural foundation is a slow pace of living around natural seasons and a strong sense of familiarity in local communities. Individuals in one village see each other every day and know everyone since childhood. This intimate feeling is developed through frequent and repeated interactions over a long period of time and within one place.

Undoubtedly, the rural foundation started to dramatically erode after Chairman Mao Zedong passed away in 1976 and Deng Xiaoping launched economic reform in 1978. The process of modernization in post-Mao China, reified through industrialization and urbanization, has jeopardized and even demolished home-related cultural norms to such an extent that ordinary people seem bereft of home. First, hundreds of millions of people have migrated from villages to cities and stopped relying on soil for livelihood;[5] second, urban residents in those same residential communities do not know each other, let alone have deep and meaningful interactions; and finally, these individuals must manage the manacles of acceleration and rationality that have entered into private life to organize their time and space (Pun 2005; Rozelle and Hell 2020; Wu, Zhang, and Webster 2014). While Fei's emphasis that Chinese society grows out of its ties to the soil reminds us of a deep rural foundation in China, a new episode unfolds in the course of speedy modernization in the post-Mao era to implicate ordinary individuals in dramatic sociospatial changes. From rural to urban, from familiarity to strangeness, from slowness to speedup, China has undertaken a great transformation from a *xiangtu* society to an unhomely one.

When a lifetime burns under the forces of speed and pressure, the home inherent in Chinese society, which cannot be deciphered through wealth, property ownership, or social status, shatters into powder in mere decades. This is exactly the suffering that Mahua endured in Beijing. Indeed, high mobilities and rapid urbanization generate "a profound sense of a loss of territorial roots, of an erosion of the cultural distinctiveness of place" (Gupta and Ferguson 1992, p. 9). The loss of home feelings has become such a cultural crisis in China that Chinese President Xi Jinping launched a nationwide campaign of *Jizhu xiangchou* (literally, remembering hometown, 记住乡愁) in 2013.[6] To President Xi, remembering hometown means to retrieve cultural roots in the countryside so that individuals do not feel homeless or uprooted. Interestingly, the experiences of homely life by my respondents in Lijiang are centered upon three elements that Fei Xiaotong highlights in Chinese *xiangtu* society: closeness to nature, enjoyment of cyclic time, and finally, socialization for a sense of community. Against this backdrop, *Unhomely Life* places the home in the context of China's speedy modernization to investigate how ordinary individuals experience a sense of home in order to mitigate the loss of home feelings in a rapidly changing society.

China's transformation since 1978 has always provoked academic and public interest, and this transformation has sped up since China joined the World Trade Organization and became integrated into the global economy in 2001. Its rise to the status of a global economic power in a relatively short period impacts every aspect of life within and beyond the Chinese territory. Less examined is how ordinary individuals in China react to such a great transformation characterized by time discipline, rising wealth, and sociospatial alienation. For instance, middle-class anxieties have become a popular theme to provoke intense debates across the Internet and in newspapers in China and elsewhere. Many residing in China's big cities engage in the merciless race for wealth, but concurrently develop a self-perpetuating sense of insecurity. Against the background of growing insecurity, discomfort, and anxiety, people attach new importance to home. In her excellent book on homemaking, *In Search of Paradise: Middle-Class Living in a Chinese Metropolis*, Li Zhang (2012) delineates the significance of home as a paradise to middle-class residents in Kunming, the capital city of Yunnan province. She points out that instead of seeking happiness and fulfillment through socialist ideals, these professionals and entrepreneurs prefer to enjoy material comfort and social distinction in newly constructed gated communities in the quest for a good life. The shift from collective sacrifice in Maoist China to conspicuous consumption in the post-Mao era redirects the meaning of home towards personal oasis, class status, and personal identity.

Zhang's work on homemaking in Kunming is valuable precisely because it provides deep insight into ubiquitous anxiety among middle-class urban residents and their struggle for refuge and oasis in their own private property. It illustrates how the rise of private homeownership is reshaping urban space and daily life in contemporary China. While I share Zhang's focus on the representations and experiences of home, I go beyond her combination of house and home and emphasize home as a subject of inquiry in its own right. By doing so, my analysis transcends the materiality of housing to specify that a homely way of life is not necessarily confined within the private sphere. Different from Zhang's focus on residents who dwell in one city, I prioritize mobility as a central entry point because mobility provokes "a new experience or orientation and disorientation, new senses of placed and placeless identity" (Morley and Robins 1995, p. 121). On the one hand, the rapidly accelerating mobility of people generates a profound sense of uprootedness and erodes the cultural distinctiveness of places (Gupta and Ferguson 1992). On the other hand, individuals make efforts to withstand a loss of home feelings by traveling for a sense of home elsewhere (rural countryside, natural resorts, parents' house, or heritage sites) or in another time (elsewhen) (nostalgia, childhood, or traditional communities).

As I explained at the beginning, Mahua and others from China's big cities have an adequate ability to make their home by means of migration or retreat-oriented tourism. But in Lijiang, many are pushed to move. For instance, native residents are socially displaced from their ancestral courtyard houses, while service workers are spatially dislocated from rural villages for tourism-related job opportunities in the town. Mobilities are essentially uneven. Bauman (1998) classifies mobile individuals into two categories: vagabonds and tourists. The former have no choice but to move, while the latter travel because they have freedom to do so. To Bauman (1998, p. 94), "tourism is the only acceptable, human form of restlessness." The point is well taken. But given that social and spatial mobility commonly transform and dominate everyday life in China and elsewhere, how do we arrive at an age of modernity characterized by homelessness and meaninglessness? When individuals are called upon not to become vagabonds, but rather to work hard, get rich, and become tourists, where is the place that people can call home? How do they strive for an ideal home elsewhere and elsewhen? Or asked differently, if modernity signifies what Karl Marx describes as "all that is solid melts into air," why do people on the move, regardless of class, gender, or age, want to hold the home as solidly as possible?

Addressing these questions, I studied four groups of people – lifestyle migrants, retreat tourists, native residents, and local service workers. The research case is Lijiang Old Town in Yunnan province. These four groups came from different parts of China to converge in Lijiang and engage in making and unmaking home in their everyday life. To be clear, the individuals I study are not houseless. Rather, they can secure a decent abode, and some even are rich enough to own multiple properties. Different from those marginal groups who suffer from a housing shortage or dilapidated housing in Chinese cities (Pun 2005; Rozelle and Hell 2020; Wang 2004; Wu et al. 2014), my respondents are not harassed by housing issues. Their unhomely life means that they engage in an endless journey to reconcile the gap between the unhappy home they leave behind or escape from, the existential home they make where they currently dwell, and the ideal home that they imagine for a better life. Filled with confusion and negotiation, their journeys epitomize efforts to strive for a relatively homely life under a given set of constraining circumstances in China.

In *Unhomely Life* I have two primary objectives. First and most importantly, I seek to explore the making and unmaking of home by individuals, in order to unravel how they handle the loss of home feelings and experience a homely way of life. The process of homemaking entails various layers of contradictions and transactions. While lifestyle migrants and tourists hope to escape speedy modernization by staying in Lijiang Old Town and enjoying an "ideal" home characterized by relaxation and loafing, native town residents give up the right to use their own courtyard houses, and local service workers are brought into the tourism industry to offer homely service to tourists. While tourists must return to big cities and reengage in speedy life after a temporary stay in Lijiang, native residents and local workers prefer to become "hometown babies" (jiaxiangbao, 家乡宝) and embed themselves in local time-space. While elderly native residents feel nostalgic about the loss of Naxi aura in the old town and insist on staying in their own courtyard houses for peace and comfort in a clamorous tourist setting, young natives swiftly move to the new city for modern lifestyles akin to those in China's big cities. While female tourists can travel alone and stay in Lijiang Old Town for new experiences, local females must let go the dream of traveling elsewhere and long-distance migration in order to shoulder the responsibility of homemaking in a patriarchal society. This list can go on, but the message is clear. Homemaking involves transactions and contradictions, revealing a synthesis of continuity and disruption to shape individuals' efforts to make home.

I argue that the home, as an embodiment of ideal existence, may never be fully identified or reached, yet it motivates individuals to launch journeys in search of a better life. Furthermore, individuals on the move are not passive subjects in China's modernity. They are simultaneously empowered and disempowered in the modernization process in post-Mao China, since a ubiquitous matrix of political-economic power is often at work, enabling their pursuit of home and pulling them back under direct control. The making of home is fraught with tension and inequality. Specifically, I highlight four types of inequality in terms of gender, age, class, and ethnicity, which can be reflected in time and space. The spatial and temporal unevenness in the process of homemaking, I argue, illuminates some of the most significant trends in contemporary China, perhaps most strikingly vis-à-vis commodification and modernization experienced by Chinese individuals in the post-Mao era. My focus on homemaking extends inquiry beyond established analysis of housing and family to encompass the variety of negotiations and transactions that deduce the loss of home feelings and stimulate a sense of home. This analysis of life journeys between homes – as a point of departure and a destination of return – remains a heuristic tool of exploration into modernity and mobilities.

Second, *Unhomely Life* uses the home as an analytic frame to exemplify China's great transformation from a poor society on the global periphery to a relatively affluent society with high industrialization and mass consumerism. But how does this great transformation, a term borrowed from Karl Polanyi who detected the dramatic shift of European civilization from the preindustrial world to the center of industrialization, descend upon individuals in China and shape their life? According to Polanyi (2001, p. 33), one immediate result was that England undertook an almost miraculous improvement on the capacity of production, an improvement accompanied by "a catastrophic dislocation of the lives of the common people." By all means, what happened in post-Mao China is analogous to the transformation undertaken in Europe since the Industrial Revolution, albeit at a more dramatic speed. One common theme between these two transformations is the loss of home feelings, an emblem of catastrophic dislocation reflecting and reinforcing political and economic power to discipline time and space for the sake of profit making. Following Brickell (2012, p. 228) who argues that home is "a vital space for understanding the micro-geographies of social and spatial uncertainty," I am concerned with comprehending the complex interplay of personal life journeys of (un)making home and the broader context of China's social and spatial changes. Hence, my analysis responds to C. Wright Mills's (1970, p. 162) call for the need to "grasp the interplay of intimate settings with their larger structural framework." After all, no setting is more intimate than the home.

A detailed analysis of homemaking can generate vital clues to understanding both the continuity and disruption in China's great transformation. In 1946, in a pioneering article on the Chinese society, Fei Xiaotong envisioned that dramatic changes would descend upon China once the country was brought into the orbit of the Western-centric global economy.

> It seems that traditional China achieved a certain equilibrium from which ensued stability. This equilibrium was upset when China came in contact with the Western powers, with their industrial supremacy. Modernization is imposed on the Chinese by the machine age, and China is forced to enter the world community. Hence the change in the Chinese social structure. (Fei 1946, p. 12)

Nevertheless, Fei did not have a chance to update his vision, although he published *Xiangtu Zhongguo* in 1947 to further explain the rural foundation of traditional China and how this ensured stability played out among villagers. *Unhomely Life* provides an update on the change in the Chinese social structure in the rapid process of modernization and urbanization, particularly since 1978.

While these two primary objectives permeate *Unhomely Life*, I also pursue a third, though less ambitious, objective. Rather than evaluate or even outline various strands of philosophical thinking about home by ancient Chinese scholars, I will deploy some of their thoughts to frame my analysis of homemaking in contemporary China. Some influential thoughts include a destination of return to external and internal nature, and a balanced schedule between busyness and loafing. This engagement with Chinese traditional culture aims not to promote Chinese exceptionalism nor to provincialize China into a case easily explained by "universal" Western theories. Rather, I concur with Hamilton (1985, p. 207) who argues that:

> China is not a mirror image of the West. Rather it is an independent vision of how a society can be put together, and must be understood in its own terms. Cross-cultural comparisons are an important means to fashion such independent understandings, but China should be treated as a positive case [and] accordingly, the concepts and the theories of analysis should take their meaning from China's historical experience.

Hamilton addresses the question of why there is no capitalism in China. While capitalism as a mode of political economy was indigenous to Europe and brought to China in the late nineteenth century, home is deeply rooted in Chinese civilization and has its own meanings in Chinese culture. Thus, *Unhomely Life* neither advocates homemaking in China as a mirror of European experiences, nor echoes Yan (2009) who argues that China presents an alternative model of social transformation in the age of globalization. Rather, *Unhomely Life* highlights the geographical and historical variants of homemaking to enrich the analysis of modernity and mobilities experienced by ordinary people in variegated socioeconomic positions in China.

Taken together, *Unhomely Life* seeks to address both theoretically and empirically unhomely life, a condition between the loss of home feelings and the experience of homely life, that exists in China under dramatic transformation. To counter what Polanyi (2001) calls "catastrophic dislocation," ordinary people make great efforts to make home, not through house ownership, but through imaginations and experiences. They render home as necessary and essential to their life because they want to assert their anchoring place in an era of speedy modernization, to achieve temporal and spatial autonomy in the midst of time discipline and spatial alienation, and to retrieve the meanings of life without succumbing to the logic of money making. The pursuit of what constitutes a homely life, both spatiotemporally and socially, is often what human agency is about.

Blending a robust theoretical framework of unhomely life with in-depth and sophisticated empirical data collected from 2002 to 2020, *Unhomely Life* offers three distinct contributions to the scholarship of China studies and critical geographies of home. First, the Chinese state mobilizes discourses of nationalism, civilization, and modernization to shape and control people's everyday life. This efficacy of a state-oriented mode of modernity is well illustrated by scholars who study the Chinese state and its relationship with society, such as Teresa Wright (2010) and Hui Wang (2009). The literature regards the Chinese state's projects of modernization as an attempt to tame and control the path and pace of modernity, and thus integrate ordinary individuals into a modern society of perfect order. But can the discussion of state-oriented modernity explain the changing everyday life experiences of individuals? The answer is no, simply because these authors neglect individuals' agency in negotiating the logics of political control and capital accumulation.

We need a more nuanced analysis of social and spatial transformation in China. *Unhomely Life* offers a significant and timely intervention for scholarship on mundane China, because Chinese society in the post-Mao era is differentiated into various social hierarchies. Yet our academic frames for analyzing such a trend of differentiation focus on political (party membership and administrative ranking) and economic (housing, urban-vs-rural, and income) variables, and thus remain deficient compared to China's actual complexity. In response to these deficiencies, *Unhomely Life* underlines how variegated social differences are entwined to reshape individuals' unhomely life. I consider the coexistence of various tensions when mobile individuals from China advance their own versions of home alongside social differences – class, gender, ethnicity, and age – that unfold unevenly in time and space.

Second, we surprisingly ignore the role that home plays in orchestrating people's life journeys in China.[7] One reason that explains such a stark contrast is that scholars of China studies have relied largely on the analysis of housing, family, or domestic life, and mixed these elements with home. For instance, the established literature on housing generates a dominant conceptualization of home as a personal or household asset for exchange or residency in China (Huang and Li 2014; Liang 2014). This conflation of housing and home explains why scholars take for granted the making of home in China, even though home is an essential component of Chinese culture. They assume that people can naturally cultivate a sense of home as long as they own real estate property in big cities. This assumption is epistemologically problematic and intellectually flawed. Meanwhile, Jie Li's (2015) book, *Shanghai Homes: Palimpsests of Private Life* is a valuable combination of micro-history and memoir to remap the private lives of the author's family and some neighbors in the alley where she once resided onto Shanghai's broader cultural landscape and cosmopolitan history. Li shows that Shanghai homes have been embodied in territories, artifacts, and gossip through which individuals struggle to absorb and adapt to major historical changes from the semicolonial period before 1949 and the communist revolution, to the reform period. Her focus on private life runs the risk of mixing home and domestic chores into one subject of academic inquiry. Departing from property ownership and domestic responsibility, *Unhomely Life* delves into the cultural construction of home: a bridge between inner nature and external world and a balanced way of life in this actual world. Through this conceptual construction, the book reveals the lived experiences of home for loafing and slowdown in a stark contrast to social acceleration and time discipline in China today.

Finally, scholars in the critical geographies of home argue for the need to exemplify the simultaneity of belonging and alienation in order to cultivate a politicized understanding of home (Blunt and Dowling 2006). The attack on home through massive demolition and demographical replacement generates an effect of horror and disorientation that implicates individuals into an uncanny condition and deprives them of their sense of home (Blunt, Ebbensgaard, and Sheringham 2021). This deprivation is particularly true for people on the margins. For instance, Brickell (2020) reveals the violence and gendered contingency of existence that women in Cambodia must endure. As one of the most idealized sites of human existence, the home offers a critical entry point for understanding and unraveling the micro-geographies of social and spatial uncertainty in a rapidly changing world. While I do recognize the devastating social and economic realities, I tell a different story by unraveling the agitated swirl of homemaking below the seemingly peaceful surface of Chinese society. The society at large is comprised of ordinary people who might not necessarily suffer affliction and houselessness as marginal groups do. Having said this, these individuals must manage the manacles of speed and rationality that have entered their daily life and made them feel not at home. Furthermore, while acknowledging the significance of home to individuals, I remain attentive to the continuing transformation of home in response to mobilities and modernity. This book advances the critical geographies of home by eschewing accounts of individuals traversing the tenuous line between here and there, the past and the present. The purpose is to unpack and analyze the in-between condition of living here and yearning for elsewhere as a common feature in a society under dramatic transformation. Next, I introduce two key terms: modernity and mobilities.

Modernity as a Deal

Modernity is a powerful logic of organizing human life, and cuts across all boundaries of geography and ethnicity, of class and nationality, of religion and ideology. Anderson's (1984) work posits that modernity entails two interrelated lines of development. First, it is embedded in economic relations that substantially transform society unleashed by the advent and entrenchment of the capitalist world market. In other words, development is essentially, but not exclusively, "*economic* development" (Anderson 1984, pp. 98, original emphasis), or to paraphrase Marx, capital accumulation for maximum profit by all means. The second line refers to self-development that momentously transforms individual life and personality, a transformation aiming for inner-freedom (Marcuse 1964), or "a heightening of human powers and widening of human experience" to mediate the logic of capital (Anderson 1984, p. 98).

Undoubtedly, economic development has received more favor than self-development. The experience of modernity, as Marshall Berman (1982, p. 15) critically asserts, is to "find ourselves in an environment that promises us adventure, power, joy, growth, transformation of ourselves and the world—and, at the same time, that threatens to destroy everything we have, everything we know, everything we are." As a result, the meanings of time and space are restructured for the sake of both capital accumulation and political control. Writing about modernity from a temporal perspective, Tomlinson (2007) argues that speed has been always associated with emancipation, creativity, and efficiency, and the cult of speed also raises concern about the loss of control over personal time. In *Modern Times, Modern Places*, Peter Conrad (1999, p. 9) succinctly asserts that "modernity is about the acceleration of time, and also the dispersal of places." Similarly, Rosa (2013) notes that in the wake of the Industrial Revolution, the feeling of speed intensified, and an acceleration of tempo instructed the practices of everyday life. Thus, Rosa further argues that the experience of modernization is an experience of acceleration: "acceleration is the central feature of the transformation of time structures, and as such a fundamental force that shapes the culture and social structure of modernity" (Rosa 2013, p. 21). This loss of time is tidily captured by Adam (1998, p. 27) who emphasizes the overwhelming role played by clock time in orchestrating everyday life: "machine time has been reified to a point where we have lost touch with other rhythms and with the multiple times of our existence." Speed and the accompanying logic of time discipline collapse temporal horizons into one, and possibly the only one, linear time for the goal of capital accumulation. Or as Tomlinson (2007, p. 53, original emphasis) asserts, what remains is "*one* single story, in which there is a continuous coupling of modernity to speed and to immediacy."

Because modernity is so universal and omnipresent, however, it generates a counter logic to modulate itself. Specifically, individuals engage in a form of countermovement so that they can discover the self and mitigate the ills of modernity. For instance, Giddens (1991, p. 20) emphasizes that one of modernity's characteristics is a relentless process of disembeddedness in which social life is propelled away from "the hold of pre-established precepts or practices" associated with established values and traditions. This disembeddedness leads to a form of countermovement, a subjective pursuit of meanings, through which modern individuals can make their life meaningful. The result, according to Anderson (1984, p. 97), is that modernity is enfolded in "a paradoxical unity, a unity of disunity: it pours us all into a maelstrom of perpetual disintegration and renewal, of struggle and contradiction, of ambiguity and anguish," giving rise to the ambivalence of modernity. In terms of tempo, modernity always carries its own paradoxical countermovement: the preponderance of slowness and stoppage as an unintended but inevitable consequence of social acceleration. Or as Rosa (2013) argues, pathological slowdown may appear as a reaction to and withdrawal from the social pressure to accelerate.

To understand the ambivalent nature of modernity, I follow Harari (2016) to succinctly define modernity as a deal made between the self and the world. The deal seems extremely complicated, since it involves social, economic, and political matters. In essence, the deal can be simply described as follows: "humans agree to give up meaning in exchange for power" (Harari 2016). This exchange offers individuals an irresistible temptation for power in its forms of money, time, technology, weapons, and so on, at the price of the meanings of life. The exercise of power for collective control and individualistic desire is underpinned by "the abyss of complete nothingness" or a widespread void of meanings (Harari 2016).

Figure 1.1 illustrates the formation of unhomely life in relation to the deal of modernity. It represents a seesawing existence – home and homelessness – in which individuals conduct journeys between the two poles of a seesaw. It can be a sense of home uprooted forcefully or involuntarily, or the revelation of homelessness at home, or some condition in between. Either home or homelessness marks an important crescendo in one rite of passage, which cannot last perennially. Then, the rite reverses the direction towards the other side. One process of homemaking has a beginning and an end, and this process transforms itself into its opposite direction, that is, home unmaking, either because the logic of capital accumulation endeavors to melt the solid into the fluid to speed up profit making, or because individuals choose to launch a new journey. Meanwhile, individuals make great efforts to dissolve a sense of homelessness, regardless of whether they are housed or unhoused, through building attachment to social or spatial entities.

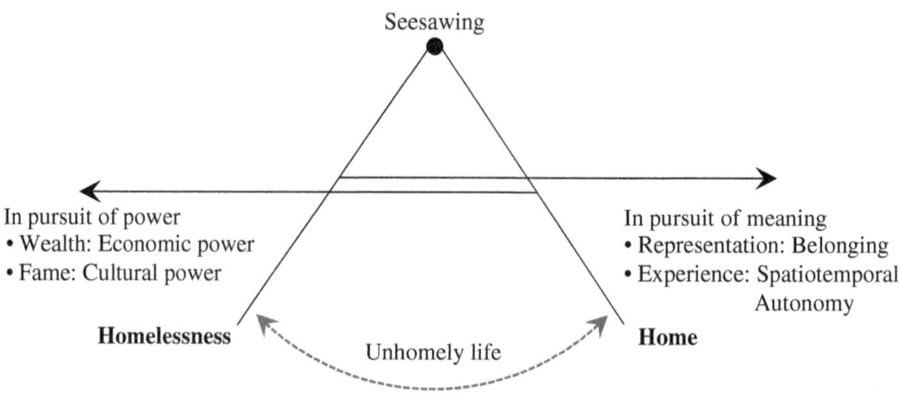

Figure 1.1 A conception of unhomely life between power and meaning.

In this sense, home and homelessness constitute two aspects that coexist in the single entity, that is, human life; furthermore, these two aspects transform into their opposite: from time to time, from here to there. The movement from one pole to the another, through different imaginations and experiences in accordance with personal agency and external conditions, fosters an endless journey of unhomely life.

The deal of modernity existed in Maoist China, once political power overwhelmed other forces to shape the meanings of individuals' lives. As a categorical symbol of state-oriented modernity, the Maoist era (1949–1976) was implicated in the endless campaigns of creative destruction. The famous slogan during the Cultural Revolution (1966–1976), "destroy the old world and build a new world," highlighted the ambition to eradicate all traces of the old for a new China. One early initiative of the Cultural Revolution was to authorize and mobilize Red Guard units in August 1966 to launch the nationwide campaign of destroying the Four Olds (old ideas, old culture, old customs, and old habits) (Clark 1996; Yang 2004). Goethe's Faust, the figure of modernity in the West, was portrayed as an epic hero to "destroy religious myths, traditional values, and customary ways of life in order to build a brave new world out of the ashes of the old" (Harvey 1989, p. 16). To some extent, Mao is a Faustian figure in China. The Maoist revolutionary project against the Four Olds did not succeed in building a new socialist China, although it afflicted plight and dismay upon tens of millions of ordinary people. Nevertheless, the modernization project advocated by Deng Xiaoping after 1978 almost smashed the olds and imposed new meanings on individuals' lives.

In 1978, Deng Xiaoping launched economic reform and called for four modernizations in the fields of agriculture, education, military, and technology. At the very beginning of the reform, progress was the most extreme manifestation of China's project of modernity, representing a promising future to stir up radical optimism among ordinary Chinese people. Since then, the logic of capital accumulation has ravaged everyday life and plundered traditional community ties, forcefully dissolving the established social norms and cultural tradition in order to pave the way for the discipline of time and space upon individuals for the sake of profit. Capital accumulation, for Marx, is indeed a dramatic destruction of the old social world. This exactly describes the current situation in post-Mao China.

The deal of modernity in post-Mao China draws its legitimacy from a version of progressivism geared towards what Deng Xiaoping calls common prosperity. In March 1985, Deng (1993, pp. 111–112) specified that "the aim of socialism is to make all our people prosperous, not to create polarization." Four decades later, however, the progress that socialism promised now stands for the threat of a relentless and inevitable transformation that does not augur peace and prosperity but rather continuous crisis and strain. Indeed, Heller argues that modernity drew its legitimacy from a version of progressivism, that is, a distant future that is allegedly riddled with great expectations and sweet honey. The result is, as Heller (1999, p. 8) notes, completely at odds with the version: "yet, instead of Heaven descending upon Earth, Earth was transformed into Hell. The fast trains ran toward their final destination—and the names of the terminal railway stations were Auschwitz and the Gulag—the stations of extermination." Echoing Heller's viewpoint, Bauman argues that the progress that modernity promises evokes "insomnia full of nightmares of 'being left behind,' of missing the train or falling out of the window from the fast-accelerating vehicle" (Bauman 2005, p. 161). In China, nightmares continue to plague ordinary individuals, although the source has been shifted from Maoist political campaigns to Dengist economic progressivism.

Today, China's journey of modernization continues with restless acceleration. Nevertheless, the sentiment that people rush to eradicate the nightmares of "being left behind" is so strong that everyone struggles to either crowd on to this high-speed train or catch up with it. To be qualified for the train, ownership of one housing property in a first-tier city (Beijing, Shanghai, Guangzhou, Shenzhen) is treated as a legitimate ticket to board, and owning two properties means good safety.[8] For those who do not own a property in these cities, they must hurry up and seize one; otherwise they lose in this knockout match and feel left behind. Indeed, property ownership in cities, particularly in first-tier cities, has become a visible form of wealth that shapes marriage, child education, and social status (Huang, Yi, and Clark 2020). Wei, Zhang, and Liu (2017) find that in Chinese cities, people pursue larger and costlier apartments with the hope of improving their status in the marriage market. Once an estate property becomes a rigid demand in order to get married (Wang 2018), more and more stay away from marriage because their earnings cannot keep pace with the skyrocketing housing prices in big cities, and those who are married dare not have babies, which aggravates China's plummeting birth rate. For those on the train, they fear they will be kicked out; for those chasing the train, they fear they will never be able to board. It seems that the whole society is imbued with a sense of fear and insecurity. The more individuals discipline their time and space for wealth and "home" ownership, the less they develop a feeling of being at home. Progress seems to have unbendingly moved to "the opposite, dystopian and fatalistic pole of anticipations" (Bauman 2005, p. 161).

By analyzing how home is spatially and temporally (un)made, *Unhomely Life* unravels how ordinary individuals contribute to and mediate with speedy modernity. Since 1978, ordinary individuals in China have been allowed to engage in the deal of modernity: trading meaning (dignity, home, and time) for power (property ownership, wealth, and fame). First, people embrace speed and efficiency to fight against monetary poverty, and when they become relatively rich, they are haunted by existential poverty because the race for wealth pushes them to leave their hometown behind and discourages them from investing time to build meaningful relationships within their new cities, giving rise to a loss of home feelings. Then they make great efforts to remake their home by changing their time-space for a homely way of life. Their unhomely life gets stuck in a situation of seesawing between power and meaning, or homelessness and home. This is the deal of modernity between power and meaning, and few can raise themselves beyond this deal.

Two Dimensions of Uneven Mobilities

As a spur to modernity, mobility entails a dynamic process of deterritorialization and reterritorialization, starting from moving out of one place and residing in another for temporary or permanent settlement. Everyday life is constantly expressed through and embedded in mobility, that is, "the free choice of place, travelling, seeing the world" (Bauman 1998, p. 121). Heller (1999, p. 7) uses mobility as a metaphor to analyze modernity:

> In the modernist view, the present is like a railway station where we denizens of the modern world need to catch one of the fast trains that run through, or stop in this location for only a few moments. Those trains will carry us to the future. Settling in the railway station would have meant stagnation—for them.

Indeed, the dawn of modernity was expressed through and celebrated by Thomas Cook's rail travel, which enabled the British working class to visit the seaside beginning in the 1850s (Su 2014). According to Butcher (2003, p. 5), the achievement of capitalist modernity has been nothing less than "the democratization of leisure travel, from the few deemed worthy, and wealthy enough to partake, to an everyday activity for the majority in developed societies."

An engagement with mobility does not mean that everyone is equally caught up in the dynamism of modernity. Mobility is inherently uneven. To Sheller (2016), uneven mobilities entail divergent pathways and differential access as a result of spatial designs, physical infrastructures, and symbolic impediments. Here, uneven mobilities with regard to the ability to move, place of choice, and settlement, precipitate an obviously gap amongst mobile individuals as they now act in relation to ever-changing economic and social conditions. They move under different circumstances and for different reasons, and have different degrees of time and space sovereignties. Thus, uneven mobilities take the form of "uneven *qualities* of experience, uneven access to *infrastructure*, uneven *materialities*, uneven *subjects* of mobility, and uneven *events* of stopping, going, passing, pausing, and waiting" (Sheller 2016, pp. 17, original emphasis). While offering an exhaustive list of mobile groups is beyond the scope of this book, I outline two dimensions of uneven mobilities.

One is proposed by Agnes Heller. In an interesting essay on what we are at home, Heller (1995) introduces two typical persons: one middle-aged man who owns a shop in Rome's Campo dei Fiori and rarely steps out of his neighborhood to touch upon other areas in the city, and one middle-aged woman who travels all over the world and regards where her cat lives as her home. If the former is firmly rooted in a geographical center and deeply committed to the neighborhood, the latter has no mooring to any specific place and becomes geographically promiscuous. If the shop owner in Rome represents the culture of a place, then the cosmopolitan lady signifies the culture of a time, that is, the absolute present. Through this comparison, Heller (1995, p. 1) notes that the woman's geographical promiscuity symbolized "something uncanny (*unheimlich*), namely the abandonment of, perhaps, the oldest tradition of the homo sapiens, privileging one, or certain, places against all the others." This experience of geographical promiscuity results in lack of orientation, which can elicit with urgency the questions of the ultimate purpose of being in the world, and whether it proceeds in a determinate fashion from our homely origin toward a homeless end. Meanwhile, Freud (2003, p. 220) emphasizes that the *unheimlich* compels individuals to search for the *heimlich* by revisiting "what is known of old and long familiar." At stake in this being known of old and long familiar is the effort to withstand a sense of disorientation. But the question is: known of and familiar by what?

Such a sense of familiarity intentionally returns again and again to one and the same spot, that is, the home. Even for the cosmopolitan woman that Heller (1995) bumped into in a jumbo plane, the home as where her cat is remains a mooring to make her develop a sense of being old and long familiar. Indeed, the cosmopolitan lifestyle that this woman experienced is as follows: "She stays in the same Hilton hotel, eats the same tuna sandwich for lunch, or, if she wishes, eats Chinese food in Paris and French food in Hong Kong. She uses the same type of fax, and telephones, and computers, watches the same films, and discusses the same kind of problems with the same kind of people. She has a 'home experience' of a kind" (Heller 1995, p. 2). Meanwhile, the shop owner in Rome's Campo dei Fiori enjoys a strong sense of old and long familiarity in his neighborhood since he prefers to center upon his daily life in a stable place. Between the local and the cosmopolitan, one thing is needed: developing what is known of old and long familiar for a sense of home. Or as Duyvendak (2011) contends, all people still can and want to attach, regardless of how mobile they are. A feeling at home requires attachable goods, ranging from Hilton hotels and familiar food to familiar neighborhoods and streetscape.

Another dimension of uneven mobilities is described by Bauman (1998). To Bauman, the spatiotemporal unattachment of nomadism in the era of classical modernity signifies a sense of backwardness and exclusion in comparison to settled life. Late modernity reverses the conception of mobilities, since the socially subordinate classes are fixed in a place and deprived of temporal autonomy. Today, immobility appears backward and left behind in the high-speed train of development. It does not necessarily mean that everyone can enjoy an equal distribution of mobility. Those who are positively privileged can stay or travel at will;

meanwhile, those who are negatively privileged have no choice but to move although they do not know where they can be welcomed to stay. Bauman calls the former *tourists* and the latter *vagabonds*. As Bauman (1998, pp. 92–93, original emphasis) puts it,

> The tourists move because they find the world within their (global) reach irresistibly *attractive*—the vagabonds move because they find the world within their (local) reach unbearably inhospitable. The tourists travel because *they want to*; the vagabonds because *they have no other bearable choices*. ... Vagabonds are travellers refused the right to turn into tourists. They are allowed neither to stay put (there is no site guaranteeing permanence, the end to undesirable mobility) nor search for a better place to be.

In the case of tourists, mobility reflects and reinforces the modern aspirations of freedom and cosmopolitanism, while to those forced to move or stay, including refugees and asylum seekers, mobility is associated with anxieties and entrapment on the edge of modernity.

If mobility represents a marker of privilege and advantage, immobility has already become "a sign of deprivation" (Bianchi 2006, p. 65). Vagabonds, who are unable to move or forced to move, strike a stark contrast to tourists who stay or move at their heart's desire. Different from tourists, vagabonds can be regarded as "flawed consumers" who fail to contribute to "the prosperity of an economy turned into a tourist industry" (Bauman 1998, p. 96). Meanwhile, touring cultures constitute a new stage of global consumerism, prevalent in Western, advanced capitalist societies (MacCannell 1976; Urry 2002; Wang 1999). Once tourism becomes an inalienable right in a globalizing world of mobile consumers, immobility portrays the other side of the same coin, simply because "one person's cosmopolitanism depends on the constitution of someone else as local" (Bell and Hollows 2007, p. 30). These vagabonds struggle for access to necessary resources or must be involuntarily displaced from one location to another, leaving them marginalized in the expansion of global consumerism. Their immobility alludes to a shadow cast by the capitalist free market upon the tourism industry. The difference between tourists and vagabonds highlights the need to comprehend the operational logic of consumerism in shaping variegated experiences of mobility. Uneven mobilities develop into a formidable sociotechnical logic to organize everyday life.

While Bauman seems to use a simplified typology, I shall argue that this abstract duality generates a stark reminder of inherent differences among mobile individuals and also gives impetus to the unequal distribution of power. What is acclaimed today as mobility is geared to tourists' pursuit of comfort, a pursuit transforming many others into vagabonds. As Bauman (1998, p. 93) reminds us, "there are no tourists without the vagabonds, and the tourists cannot be let free without tying down the vagabonds." Together, they symbolize the two essential aspects of uneven mobilities. As subjects of modernity, both tourists and vagabonds encounter a similar question: Where is a place to call home? Bauman's (1998, p. 98) answer is very simple: both "become wanderers and put the bitter-sweet dreams of homesickness above the comforts of home." Departing from Bauman's nihilistic conception of dreams of homesickness, I will bring attention to the efforts made by individuals to represent and experience home so that they can renegotiate the deal of modernity in today's China.

These two dimensions of uneven mobilities, as identified by Heller and Bauman, highlight the complexity of mobilities in an era of liquid modernity. Table 1.1 synthesizes these two dimensions. While groups of high mobility have more capacity to travel at will, those in the category of low mobility face impediments to moving. Together, these four types correspond to the four groups of people I study in subsequent chapters, lifestyle migrants (cell 1), retreat tourists (cell 2), town natives (cell 3), and hometown babies (cell 4), although this correspondence is more analytic than actual.

The booming tourism industry in China over the recent decades epitomizes the rapid accumulation of wealth by middle-class citizens. For instance, rich Chinese tourists wield their buying power and discretionary time to visit foreign countries such as Thailand and Italy. Many go abroad to purchase foreign luxury products and enjoy high-end foreign service in pursuit of status. According to Arlt (2013), well-off Chinese citizens are motivated to pursue overseas travel to increase self-esteem and gain social capital and personal prestige. Ever since 2012, China has outpaced the United States as the number one country in terms of international tourism expenditure. The outbreak of the COVID-19 pandemic in 2020 abruptly suspended Chinese tourists' international shopping. Tourism has emerged as a way of privileged life for those who get rich first and fast in China, and meanwhile, those who lag behind in the train of economic advancement degenerate into vagabonds, including 300 million rural-to-urban migrants or floating populations (Chan 2010) and millions of houseless people such as rough sleepers and beggars in cities (Hao, Zhu, and Thompson 2022). Recent debates on new urban poverty in Chinese cities spotlight alarming stratification in China (Liu, Wu, and He 2008).

A critical understanding of mobilities in China reveals that, in addition to the well-documented flows of rural-to-urban labor migrants, there are increasing numbers of

Table 1.1 Dimensions of uneven mobilities.

	Style of Life	Style of Will
High Mobility	1. The Cosmopolitan: transient or no attachment to particular places; *Home as everywhere*	2. Tourist: chronically mobile for alternative experiences; *Home as elsewhere*
Low Mobility	4. The Local: strong attachment to particular places; *Home as rooted in time and space*	3. Vagabond: forced or pushed to move; *Home as lost in transit*

migrants who move from the rich coast to the poor western hinterland, from cities to countryside, from professional employments to leisure-oriented self-employment. This sort of reverse migration is spurred by the belief that they can find a better way of life and thus experience a different lifestyle. The insistence on a favorite lifestyle runs counter to the mainstream trend of migration in China, indicating an important aspect of social transformation emerging out of rising purchasing power and more freedom to move. Addressing uneven mobilities in China, *Unhomely Life* discusses various mobilities, including lifestyle migration, retreat tourism, in-situ displacement, and dislocation. These mobilities take the form of uneven access to spatial resources such as infrastructure, landscapes, and services, as well as uneven moments of staying, departing, and waiting (Sheller 2016). Uneven mobilities reveal divergent pathways and precarious access at work to shape individuals' experiences of homemaking in China. I take these conceptual tenets to specify the situatedness of mobility and the performativity of social and economic ordering.

Lijiang Old Town: The Case

Located in northern Yunnan, Lijiang Old Town is nestled in a mountainous area and stretches across Lijiang Valley 2,400 m above sea level (Figure 1.2). It is also called Dayan Old Town. The town has been the home of the Naxi ethnic group for over 800 years. Its construction allegedly started from the late Song Dynasty (960–1279) and evolved into a regional commercial center in the Qing Dynasty (1636–1912). During the early *Guangxu* period (1875–1908), the town accommodated 900 households or about 3,000 residents (Yunnan Committee for Five Series of Books on Ethnic Issues 1983). Commerce in Lijiang Old Town continued to develop after the collapse of the Qing Dynasty in 1912. It reached a peak during the 1937 to 1945 China–Japan war when the Japanese army controlled the majority of Chinese land and blocked China's imports and exports. That caused the emergence of Sino-Indian trade as an important channel to supply daily necessities to Kunming and other cities in Yunnan. The trade was carried on horseback and the main commodity was tea. This commercial link between Yunnan and Tibet became known as the Tea Horse Road (*chama gudao*, 茶马古道). Lijiang was a major trading hub where traders from both sides, Yunnan and Tibet, would conduct business. As the town's center, Sifang Square was both the market and a staging point for caravans on their way to Tibet. Peter Goullart, a Russian citizen who spent eight years in Lijiang Old Town from 1941 to 1949, describes Lijiang as a "little-known and all but forgotten, ancient Nakhi [Naxi] Kingdom of south-west China."

> I had always dreamed of finding, and living in the beautiful place, shut off from the world by its great mountains, which years later James Hilton conceived in his novel Lost Horizon. His hero found his "Shangri La" by accident. I found mine, by design and perseverance, in Likiang [Lijiang]. (Goullart 1955, pp. 217–218)

Figure 1.2 The location of Lijiang in China. **Source:** Xiaobo Su (Book Author).
Note: Cities mentioned in this book are highlighted in bold and marked in the map.

The Shangri-la mythology simulated in Goullart's dream has generated far-reaching resonance with millions of international and domestic tourists since Lijiang was opened to the global tourism market in the late 1990s.[9]

Today, the old town is territorially located in the Old Town District, an administrative unit of Lijiang Municipality, which is a prefecture-level municipality in Yunnan province. Lijiang Municipality encompasses the District and four counties (Yulong, Ninglang, Huaping, and Yongsheng). The District is the mainly urbanized area in Lijiang Municipality, an area including both the old town and the new city (Figure 1.3). Like other Chinese cities, the new city in Lijiang contains urban villages where farmers relegate their farming land to the municipal government, but retain their rural household and homestead so that they can build courtyard houses on their own. Most of these landless farmers participate in the tourism industry. Lijiang Heritage Authority, the agency with the exclusive power to manage Lijiang Old Town, is a dispatched administrative organ of Lijiang municipal government.[10]

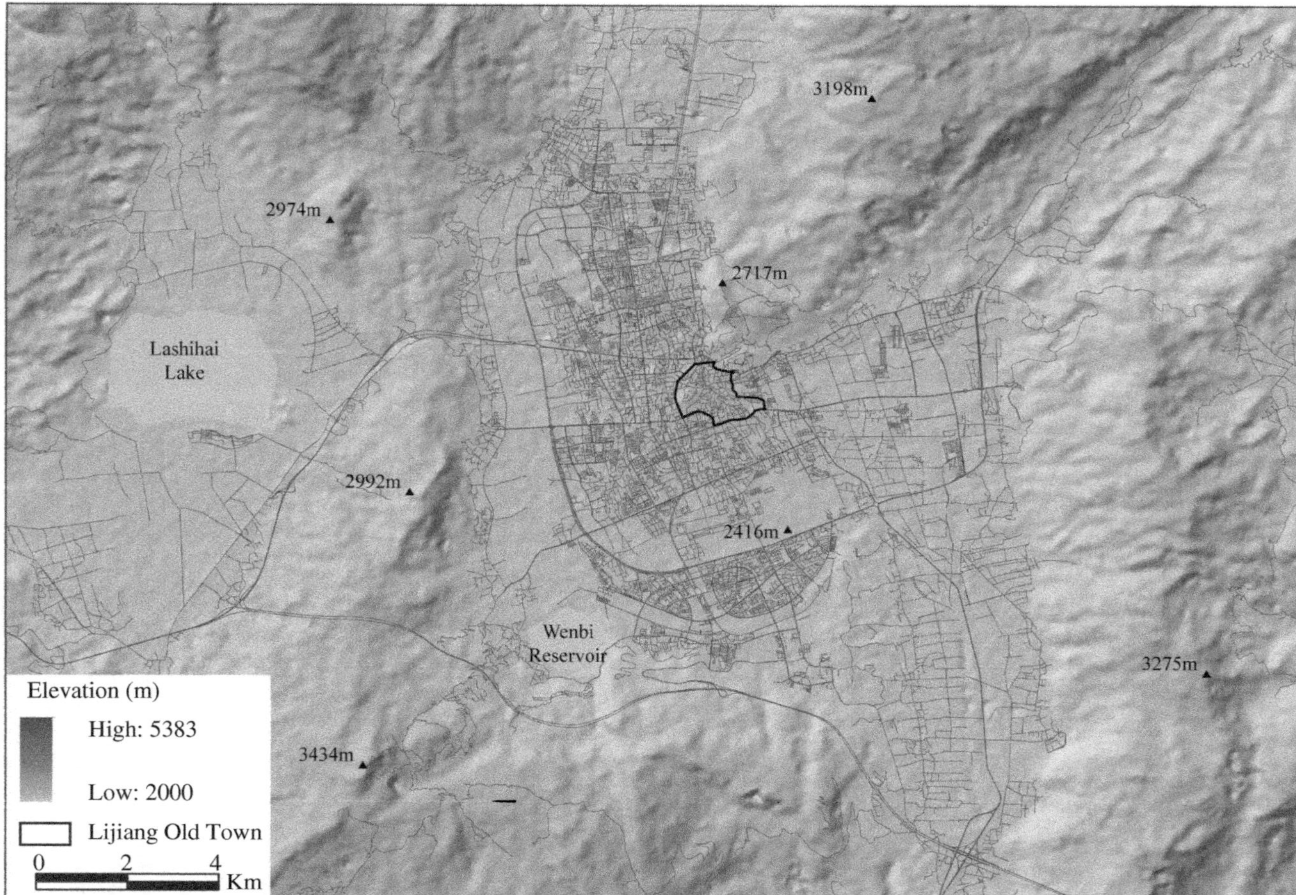

Figure 1.3 Lijiang Old Town in Lijiang Valley. **Source:** Xiaobo Su (Book Author). http://www.gscloud.cn/sources/accessdata/310?pid=302 (last accessed 20 July 2023).

The town has managed to retain its own coherence and rootedness that makes it different from many other Chinese cities where local culture and landscapes have been affected by modernization, wars, catastrophes, or other factors (Figure 1.3). Several streets, radiating from the town center, Sifang Square, connect various small alleys into a comfortable streetscape for people to stroll and chat. Sifang Square not only organized the town's transportation, but served as a daily market for the town residents and villagers to exchange goods until the mid-1990s (see Figure 1.4). In addition, the town has a remarkable water supply system that includes Yu River, West River, and East River that originate at Black Dragon Lake located 1 km north of the town. Water is channeled through several meandering tributaries, yielding a complex network of water supply in the town. In the past, the water system sustained the locals' everyday life, beautified the townscape, and brought a natural environment into Naxi people's everyday life. The alignment of streets to canals renders Lijiang a unique spatial fabric, which contrasts sharply with the rigid grid-like layout in many other old cities in China. Because of its unique urban form and authentic lived culture, Lijiang was placed on the World Heritage List in 1997. The town's core area, under strict preservation, has a size of around 1.45 square km.

After 1997, Lijiang's tourism development started to accelerate under the guise of economic growth. The town's heritage landscapes have been shaped to cater for tourists' desire for pleasure and entertainment, with bars, tourist services, souvenir selling, grand-scale performances, and ubiquitous heritage symbols. At first these changes were limited to Sifang Square and the surrounding streets most visited by tourists. In 2004, though, the Lijiang Heritage Authority decided to enlarge Lijiang's tourist capacity by encouraging and guiding tourists to other streets. The authority actually pushed tour guides to lead their customers to some sites located away from the center. Accordingly, property values rose all over the town, and the demographic displacement gradually spread from the center to every corner of the town (Su and Teo 2009). Native residents now rent their courtyard houses to migrant businesspersons who can afford the highest rent, no matter how these houses are renovated and used.

By 2008, Lijiang was no longer a residential town. It had morphed into a tourist destination with thousands of shops, cafés, and restaurants (Figure 1.5). In a sense, Lijiang became an outdoor shopping mall decorated with heritage symbols and Naxi ethnic culture. Though the Lijiang Heritage Authority implements historic preservation laws and regulations to prevent developers and businesspersons from destroying the physical fabric of traditional courtyard houses, the local character that native residents regarded as "authentic" has been overwhelmed by tourism businesses. In this process, many artists and cultural entrepreneurs, who incorporated Naxi ethnic culture into the tourism industry in the 2000s, have been displaced because they cannot afford the skyrocketing house rents.

Meanwhile, a new business movement emerged along with the commodification of Naxi heritage to cater for tourists. Lijiang lifestyle became an attractive selling

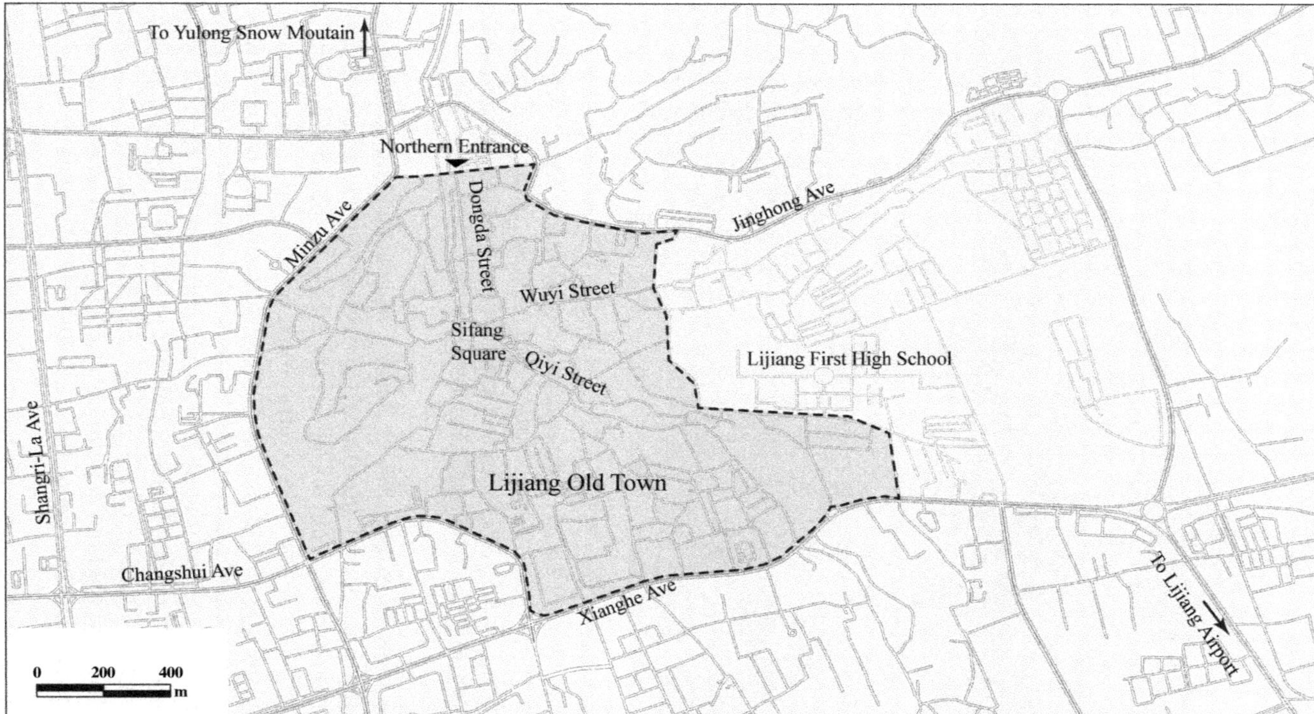

Figure 1.4 A map of Lijiang Old Town.

Figure 1.5 Sifang Square in Lijiang Old Town in 2004. The background in this photo is Lion Hill where a panoramic view of the whole old town is available. In 2012, the hill top was leveled to build Aman Dayan, a luxury hotel that opened in 2015 and offers "a serene retreat from which to explore Yunnan's natural and cultural heritage" (https://www.aman.com/resorts/amandayan). As of May 2023, the Courtyard Suite in this luxury hotel costs about 5,300 yuan (US$ 750) per night, excluding taxes and fees. **Source:** Xiaobo Su (Book Author).

point in the tourism market. Water flowing in canals, along with stone-paved alleys and the presence of traditional courtyard houses, arguably activates a sense of nostalgia among tourists who yearn for a slow-paced, tradition-oriented lifestyle. In response to speedy modernization, Lijiang Old Town can nourish the imaginations and practices of an ideal home since its pleasant natural environment reinforces its attractiveness to those tourists who attempt to escape from an unsettled and disoriented way of life that springs from the race for wealth and fame in China's big cities. The town bolsters a cultural milieu where people can gather to develop a feeling of community based on a shared lifestyle. As Zhu (2018) and Sun and Xu (2017) point out, Lijiang is an ideal place of indulgence, autonomy, pleasure, and relaxation for lifestyle migrants and like-minded tourists.

Tourism remains the most important industry in Lijiang. To illustrate, I use statistical data released by the Old Town District's government to highlight the dramatic tourism development contributed by the old town.[11] The number of tourist arrivals in 2002 was 2.68 million jumping to 3.3 million in 2005 (Bureau of Statistics in Lijiang Old Town District 2006). In the following 10 years, tourism development achieved substantial growth, as shown by 23.01 million tourist arrivals in 2015. The number rose to a record 35.15 million in 2019, but plummeted to 18.91 million in 2020 due to the COVID-19 pandemic (Bureau of Statistics in Lijiang Old Town District 2021). In the 2000s, overseas tourists accounted for 5 to 10% of Lijiang's tourism market, with a decline in the 2010s. By 2019, one year before the COVID-19 pandemic, the percentage of overseas tourists dropped to 2.83%, indicating that domestic tourists play an overwhelmingly major role in Lijiang's tourism development. Most domestic tourists come from rich coastal cities such as Guangzhou, Shenzhen, Shanghai, and Beijing, or big cities in the hinterland such as Xi'an, Chongqing, and Chengdu. In 2019, the overall revenue of the tourism industry in the District amounted to CNY 80.49 billion (~US$ 11.87 billion).[12] The added value of the service-based third sector, in which the tourism industry played a key role, was CNY 13.16 billion (~US$ 1.94 billion), accounting for 74.36% of the District's gross domestic product (Bureau of Statistics in Lijiang Old Town District 2021).

The old town's high attractiveness in China's tourism market results in more than 2,000 shops clustering along main streets and small alleys to provide services, souvenirs, and food to tourists. The phenomenal growth of Lijiang's tourism industry has led to the arrival of migrant businesspersons plopping themselves down in residential houses and running tourism businesses. At the end of 2020, Lijiang Old Town District reported that the household registered population was 161,707, among whom Naxi people accounted for 58.37% and Bai at 13.86%.[13] The seventh national census shows that by 1 November 2020, the District housed 288,787 regular residents (changzhu renkou, 常住人口), among whom Naxi people's percentage dropped to 35.87%. This demographic discrepancy indicates that over 100,000 regular residents have migrated from other cities to the District and most of them are Han people, the ethnic majority in China (Table 1.2). While most Han migrants from other Chinese cities come to Lijiang mainly for tourism-related commercial opportunities, thousands of them run tourism businesses and prioritize the enjoyment of a lifestyle characterized by relaxation and simplicity. The booming tourism industry in the town also attracts local rural villagers in Lijiang Municipality for service-related jobs in restaurants, guesthouses, bars, express delivery, and shops.

The process of demographic replacement by migrants in Lijiang Old Town attunes to gentrification. Nevertheless, the word gentrification might be too narrow to describe the profound demographic and cultural transformation that has reshaped the town. In-movement by Han migrant businesspersons and domestic tourists, coupled with native town residents' out-movement, generates an uncanny condition of in-situ displacement to those still living inside. In addition to invented Naxi ethnic heritage symbols, Lijiang's new

Table 1.2 The population of Old Town District in Lijiang in 2020.

	Household Registered Population	Regular Residents
Total	161,707	288,787
Geography		
Urban	108,553	238,828
Rural	53,154	49,959
Ethnicity		
Naxi	94,384	103,337
Bai	22,408	32,732
Han	34,106	122,247

Source: Adapted from Bureau of Statistics in Lijiang Old Town District, 2021.

"authenticity" builds upon a relaxed lifestyle in contrast with China's speedy modernization, gearing towards what Pine and Gilmore (1998) call the "experience economy," which entails four realms: entertainment, education, esthetics, and escape.

The story of Lijiang is a story of China's peripheral regions opened up to the global network of production and consumption. Under the auspices of state-subsidized infrastructure to ameliorate the distance between hinterland China and the heart of global capitalism, this opening enables individuals to travel from big Chinese cities to pursue a sense of home. In less than three decades, Lijiang Old Town has been transformed from a peripheral and dilapidated small town on the brink of being demolished for real estate development to a highly popular tourist destination (Su and Teo 2009). It has morphed into a hybrid zone that mixes features of big cities and *xiangtu* communities. On the one hand, all these mobile individuals – tourists, migrants, displaced natives, and dislocated service workers – meet each other in the town, at one time or another, to facilitate the fulfillment of a home-based experience economy that entails good infrastructure, beautified townscape, relaxation, and stories. Under the surface of this economy lie economic exploitation, demographic displacement, social differences, and uneven mobilities. On the other hand, the essential tempo inherited from traditional Naxi society, characterized by loafing and slowness, has been maintained in the town, as contemporary dwellers – migrants and retreat tourists from Chinese big cities – prefer to enjoy it for an experience of temporal autonomy. Lijiang's dramatic change in three decades is a microcosm of China's own great transformation following 1978 – the former driven by mass consumerism, the latter by industrialization and globalization. The latter lays the foundation for the former. The analogy shown in these two transformations at different scales justifies the broad value of studying homemaking by various groups of people in Lijiang.

Structure of the Book

Unhomely Life is organized as follows. Chapter 2 draws a temporal perspective to articulate a sense of home in traditional Chinese culture and how this sense has been eroded by post-Mao modernization. Rather than assuming the home as a physical shelter for dwelling, I examine two key meanings of home in traditional Chinese culture – a destination for return and a homely way of living – through ancient Chinese poetry and philosophical ideas. Then I discuss how the campaign to get rich first and fast in the post-Mao era has diluted these meanings and implicated individuals in time discipline in China's first-tier cities such as Beijing and Shanghai. To understand the loss of home and the search for home by ordinary individuals in China, I elaborate Homi Bhabha's (1994) work on an unhomely world to establish an analytic framework to understand unhomely life as a seesaw between the loss of home and the search for home. Then I propose a journey-oriented perspective: leaving a previous home to explore possibilities elsewhere and elsewhen, dwelling with belonging and alienation in the current home, and imagining a new home in the future.

The subsequent four empirical chapters respectively examine four groups of people on the move. Chapter 3 analyzes middle-class lifestyle migrants from big cities to Lijiang Old Town. These migrants' efforts to remake their home in Lijiang entail a critique of, and a compromise with, speedy modernity. By representing the town as an ideal home and a place to enjoy a homely way of life, these migrants attempt to mitigate the loss of home feelings. Then I explore the spatial politics of belonging and alienation that places lifestyle migrants in an unhomely condition. The COVID-19 outbreak in early 2020 jeopardized Lijiang's tourism development, and competition for customers intensified. Within this context, the divergence between those who resume hard and busy work and those who continue a relaxed lifestyle illustrates the variegated ways of handling the race for wealth.

Chapter 4 draws on a temporal perspective to investigate how middle-class domestic tourists enjoy relaxation and simplicity in Lijiang Old Town. As the benchmark of modernity in today's China, these middle-class tourists feel bored with their own hectic life and travel to Lijiang for a temporary retreat. The goal is to retrieve a renewed sense of home in Lijiang so that they can recharge both body and mind for the continuing race for wealth in China's big cities. I mainly focus on how they spend their time in the town and how the temporality in Lijiang is different from that offered by China's big cities. Specifically I focus on two sets of banal activities – solitude and socialization – to highlight how these activities contribute to the experience of a homely way of life. The swing between loafing in Lijiang and busyness in big cities represents a manageable and even palpable way of living with speedy modernization. The representation and consumption of home generates intensely contradictory configurations of struggle that simultaneously push tourists toward an ideal of home, yet pull them back into the whirling vortex of "modern" life and commercial forces.

If lifestyle migrants and domestic tourists represent the relatively privileged aspect of mobility in China, Chapter 5 turns to native residents to paint a complicated picture of voluntary displacement. When their birthplace degenerated into a tourist town, native residents became strangers in a familiar place. Drawing on narratives from dozens of native residents, I emphasize the uncanny feeling that developed towards the town when their familiar place became a Disney-like playground for strangers coming from other parts of China. Thus, most native residents choose to lease their properties to migrant businesspersons and then move to the new city. The contrast between the determined young natives and the reluctant old highlights the age difference in the process of unmaking home. Meanwhile, a few still choose to stay in the town and engage in tourism business related to Naxi culture. I tell four different stories of how native residents adjust to the old town's dramatic transformation, and how they negotiate the deal between money and meaning. This chapter challenges the assumption that native residents are vulnerable to external intrusion, or victimized in the process of tourism development. We cannot criticize them for prioritizing economic return over cultural preservation. Rather, I emphasize that they actively adjust to Lijiang's dramatic transformation, although between the old town and the new city they become unhomely forever.

At the bottom of Lijiang's mobility ladder are service workers from rural villages in Lijiang municipality. They define themselves as local people, in line with their household registration located in Lijiang, but work hard to make the old town a home for tourists and lifestyle migrants. Culturally, they call themselves *jiaxiangbao* (literally, hometown babies, 家乡宝). Chapter 6 studies local service workers and their ambivalent experiences of homemaking. I compare those from distant villages with those close to the old town to exemplify two different struggles of homemaking. While those close to the old town make efforts to maintain their established patterns of time and space, those from afar struggle between staying in urban Lijiang and returning to the rural countryside. Furthermore, I explore how Lijiang's patriarchal relations turn homemaking into a barrier to curb local females' ability to travel or migrate at will. Gender discrimination in Lijiang's traditional rural society deprives local females of the opportunity to travel afar and fixes them into a local home imbued with patriarchal power. To embrace the identity of hometown babies, Lijiang locals stick to established rhythms in the valley and maintain social solidarity to mediate the intrusion of modernization, but social and economic factors such as high housing prices, urbanization, and Hanization pose real challenges to their sense of home.

By way of conclusion, Chapter 7 revisits unhomely life to emphasize a permanent paradoxical conundrum of belonging and alienation in modern society, a conundrum motivating people to develop a sense of home via traveling or migration. Rather than taking home to be a self-evident totality,

I emphasize the necessity of understanding homemaking as an ongoing journey shaped by a range of social, economic, and political forces. The chapter provides a reflection on the campaign of remembering hometown, as proclaimed by President Xi Jinping in 2015, to highlight how the ideologies of home become implicated in the practices of everyday life in China today. To conclude, I cite two great scholars, Zhuangzhi and Immanuel Kant, to place home in the conjunction between inner nature and external nature, showing that individuals must figure out a position between home and homelessness in order to handle the deal of modernity within the unhomely condition of the modern capitalist world.

Notes

1 The term "native" in this book does not denote specific cultural value as in an anthropologic sense, but refers to those who were born in Lijiang and grew up in Lijiang Old Town. Due to the old town's small population (15,279 residents in 1986 and 14,477 residents in 1999), native residents created a closely linked web of integrated social relations built upon common language, ethnicity, and familiarity (Su and Teo 2009). The town as their home signified a firm cultural foundation of local Naxi society.

2 High-speed trains in China are those running at 350 km per hour in daily operation, a speed faster than that in other countries.

3 Confucius said, "While one's parents are alive, one should not travel to distant places. If it is necessary to travel, there should be a definite direction" (子曰:"父母在,不远游,游必有方"). This comes from *The Analects of Confucius*, Book IV, Liren.

4 *Xiang* means "countryside" or "rural village," and *tu* means "earth" or "soil." Together, *xiangtu* refers to "native soil or home village."

5 According to the seventh general census, by 2020 the number of residents in cities and towns was 901.99 million, accounting for 63.89% of the whole population (http://www.stats.gov.cn/tjsj/zxfb/202105/t20210510_1817176.html). The number in 1982 was 206.59 million, which indicated an urbanization rate of 20.49% at that time (http://www.stats.gov.cn/tjsj/tjgb/rkpcgb/qgrkpcgb/200204/t20020404_30318.html).

6 The character *xiang* used by President Xi is the same as Fei Xiaotong's *xiang*. *Chou* (愁) means nostalgia or homesickness. The difference between nostalgia and homesickness is explained in Chapter 2.

7 A few exceptions include Zhang (2012), Li (2015), and LaCouture (2021).

8 In early 2021, average property prices for residential apartments (100 square meters, three bedrooms) in these first-tier cities ranged from CNY 5 to 10 million (USD$ 770,000–1.55 million).

9 A detailed account of Lijiang Old Town's history can be found in Su and Teo (2009).

10 The full name of Lijiang Heritage Authority is World Cultural Heritage Old Town of Lijiang Protection and Management Bureau (世界文化遗产丽江古城保护管理局,

shijie wenhua yichang Lijiang gucheng baohu guanli ju). To simplify the expression, I use Lijiang Heritage Authority in this book.

11 Technically, Lijiang Old Town is defined as a tourist destination, but Lijiang Heritage Authority never succeeded in installing gates for admission tickets. It is difficult to calculate the exact number of tourists visiting the old town. In March 2020, a system of face recognition was installed at the old town's entrances as a COVID-19 control mechanism.

12 One US dollar (US$) is about 6.78 Chinese yuan (CNY) in August 2022. I use this exchange rate throughout the whole book.

13 Both Naxi (纳西) and Bai(白) are defined as ethnic minority groups in China. In Lijiang, Naxi is the main ethnic group by household registration.

Chapter Two
A Sense of Home in China: Then and Now

Home, from ancient times to today, is indeed the source and origin of great literature and thought, of human understanding in the world, of innovative achievements in survival and prosperity. From Heidegger to Freud, from Confucius to Qian Zhongshu, the home denotes a universal metaphor through which great scholars engage significant philosophical inquiries about being in the world and the meanings of life. Where is a place called home? Or what is our home? To ask such questions is to challenge the assumption that we have already secured a place as the home and enjoy a settled way of life in the world. Regardless of our differences in gender, class, race, age, and citizenship, the home remains the crucible of imaginations and experiences. Indeed, the home comes to represent foundational values and feelings, including community, belonging, familiarity, comfort, self-development, and so on. People dwell somewhere and sometime at home, a way of being in the world beyond simply owning a house or other shelters. Everyone wants to feel at home and enjoy a homely life, and no one wants to become uprooted into a homeless condition.

Blunt and Dowling (2006, p. 2, original emphasis) emphasize the need to understand home beyond material elements such as bricks and mortar, and to consider it as an ensemble of feelings. They argue that "home … is a *place*, a site in which we live. But, more than this, home is also an idea and an imaginary that is imbued with feelings." Thus, home is culturally constructed with subjective feelings. Drawing on Blunt and Dowling, I argue that it is necessary to separate home from both house and family, and treat the home as a research agenda on its own. Such a separation involves ways of transcending the material landscape of house and the social sphere of domestic responsibility to understand the cultural

Unhomely Life: Modernity, Mobilities, and the Making of Home in China, First Edition. Xiaobo Su.

construction of home, that is, how we feel and live. I specify two aspects of home – representational and experiential – to recognize home as an ensemble of feelings and the living form of daily life.

Additionally, this chapter explains some key modalities in Chinese cultural traditions to explain the nature of home in ancient China. Using these modalities as a background, I then delineate the dramatic transformation that happened to the social and economic construction of home in the post-Mao era. Instead of taking home for granted, I recognize the need to deal with homemaking in its spatial, temporal, and social expression – always an integral and contradictory part of human experience, if largely ignored because speedy modernity generates an unsettling effect upon a sense of home in China. The process of making and unmaking home enables individuals to make sense of, and to live with, an otherwise messy world of acceleration and alienation.

Home: An Ensemble of Representations and Experiences

At the outset I emphasize that we must refrain from deploying home as a neutral term for dwelling, settlement, or other social practices, and move instead toward a nuanced and deep analysis of imaginations, experiences, and even ideals – one that has played a central role in shaping mundane practices and mobilizing efforts for a meaningful and better way of life.[1] Home bears two connotations. First, it is a representation of positive meanings cultivated through an attachment to cultural and natural roots. Among various meanings, familiarity should be singled out. To Duyvendak (2011, p. 34), "familiarity is a necessary, though not a sufficient, condition to feel at home." Such a condition is embodied in specific entities that can be a family, a community, a building, a landscape, a city, or a nation, in line with personal feelings, social relations, and cultural bonds developed over time, where individuals can feel at home (Blunt and Dowling 2006; Duyvendak 2011). In other words, it takes time to feel at home, and the cultivation of belonging to an entity requires the investment of time, gradually sedimented layer by layer, to become familiar. Without such investment, individuals might not feel at home. This representation highlights home as a symbolic anchor of continuity and familiarity that sits in contrast to increasingly social alienation and rampant consumerism.

The relationship between nostalgia and home deserves elaboration, particularly when I address the need to employ a temporal perspective to tease out feelings of home and homelessness. While nostalgia refers to a longing for another time, a good moment of the past such as childhood, homesickness means a longing for another place in order to physically or imaginatively escape the current place imbued with dissatisfaction or disruptions (Duyvendak 2011). Both nostalgia and homesickness represent a sense of mourning over changes that have taken place here and now, particularly when the familiar turns strange and people feel alienated in their regular setting. To Rapport and Dawson (1998), the homeless mind as a result of modernization is hard to bear. This unbearable condition is particularly explicit to migrants. To counter this mind, there is widespread nostalgia for "a condition of being 'at home' in society, with oneself and with the universe: for homes of the past that were socially homogeneous, communal, peaceful, safe and secure" (Rapport and Dawson 1998, p. 31). Hence, nostalgia for the past and longing (or homesickness) for elsewhere are logically compatible to shape the making and unmaking of home: what to remember, where to go, and how to settle down for a while (Tinsley 2020). From a temporal perspective, Tinsley (2020, p. 2331) contends that "nostalgia is a disenchantment with one's lived experience of reality, and a longing for an idealized, imagined 'home.' That idea of home, further, is deterritorialized, drawn from past and present, and projected onto an imagined future--whether or not the nostalgic acts to bring it to fruition." This argument can help us to understand that home is not a fixed entity that stands still, but entails complex and even ambivalent feelings in an endless journey from the past to the present and then to the future.

Second, home is more than a representation; it is practically grounded in daily experiences in various geographical settings. A homely way of life means an assemblage of daily experiences characterized by comfort and relaxation. From the perspective of time-space, these experiences point to spatial and temporal autonomy in these settings. The term "autonomy" derives from the Greek *auto* ("self") and *nomos* ("rule" or "law"). Taken together, being autonomous means acting within a framework of rules one sets for oneself (Oshana 2006). The highest degree of personal autonomy is summarized by Confucius who explains that "At seventy, I could follow the dictates of my own heart; for what I desired no longer overstepped the boundaries of right."[2] At this stage, individuals act on their own volition to live their lives.[3] Spatial and temporal autonomy means that individuals stand at the helm to manage their space and time within certain conditions (May 1994). Therefore, the experience of homely life enables "the development of a sense of self-worth and humanity partially autonomous from dominating, exploiting, commercial or bureaucratic social structures" (Young 2005, p. 149). The experiences of home are important because they connect feelings with a cosmic grasp of larger social and economic forces that shape daily life. Nevertheless, the distinction between home as representational and experiential cannot be strictly drawn (Blunt and Dowling 2006; Brickell 2020).

Home is related to house, but any effort to conflate them causes much trouble to understand the cultural construction of home.[4] If house (fangzi, 房子) is a material asset for commercial exchange or daily dwelling, home (jia, 家) entails emotional attachment and complex feelings that can be embodied in, but definitely go beyond, spatial attributes of dwelling. House epitomizes a built entity of shelter in which people can dwell and this shelter has various formats, ranging from high-rise apartments to courtyard houses, rented bedrooms, duplexes, and so on. The materiality of house entails ownership, material furnishing and decoration, and economic value. Without time and feeling crystalized in space, a house cannot be equated with a home (Ginsberg 1999). Recognizing the difference between house and home brings into play the relations of economic and social operations that shape individuals' life journeys.

In the popular media, house and home are often mixed as a marketing strategy of selling real estate property to promote "home" ownership and certain political values. Highlighting home ownership instead of house ownership in advanced capitalist countries has further conflated house as a property of social status and commercial asset, home as a way of life, and family as an ensemble of domestic responsibility into one ideological agenda for political control and capital accumulation (Mallett 2004).[5] For instance, Margaret Thatcher delivered a speech to the National Housebuilding Council in December 1984:

> A house is most people's biggest asset. It is a large investment, and it needs protection. ... But a house is more than this. It is a symbol of security, and a stake in the future. People who own houses do so not just for themselves, but for their children. They do so as members of a responsible society – proud of the heritage derived from the past, glad to care for it, and eager to give the next generation a bit of capital to give them a start. I believe in home ownership because I believe in individual responsibility.[6]

Such a dominant narrative reinforces the ideological notion of home as a privatized property in capitalist societies. As the flag bearer of neoliberalism, Thatcher promoted home ownership in the United Kingdom to highlight the political discourse of responsible society constituted by property owners. If home ownership equated to individual responsibility, those who owned no property failed to fulfill their responsibility. Similarly, at the ceremony of signing the American Dream Downpayment Act in 2003, US President Bush noted that "Our government is supporting homeownership because it is good for America, it is good for our families, it is good for our economy."[7]

Statements like "home ownership for responsible individuals" or "the American dream of homeownership" attempt to utilize the linguistically positive connotations of home to boost the real estate market, while twisting the meaning of "home" to be either fixed residence or property value. Such a discourse on home ownership, or more accurately, house ownership, strengthens the structural forces that lead to the race for wealth and fame. Self-responsibility, which is an essential feature of neoliberalism, is reified through house ownership, but has unsurprisingly enmeshed many people into an unhomely journey imbued with middle-class anxiety and social insomnia.

Against this background, Saunders and Williams (1988, p. 88) contend that home in the Anglo-Saxon culture is represented by a house as a physical ideal and ownership as a legal ideal. In a recent monograph, Chambers (2020) analyzes how home (or more accurately, private house) has been idealized in the West as a middle-class haven and managed as an investment to signify social status and express personal identity. Even the academic journal *Home Cultures*, which is designed to study the cultures of home, proclaims its dedication to "the critical understanding of the domestic sphere, its artifacts, spaces and relations, across timeframes and cultures."[8] The conflation of house and home cannot explain why so many people cannot feel at home even though they are housed or dwell in a decent setting, and why some others are not appropriately housed, but feel at home in a neighborhood, a temporary shelter, or on the road. This conflation raises much critique in the literature. For instance, Mallett (2004) criticizes architects and historians for uncritically conflating house and home and asserting that the spatial organization of domestic dwelling in a house influences and reflects social norms and cultural values.

If a house is where we live, home means how we live (Ginsberg 1999). Recognizing the difference between house and home brings into play the relations of economic and social operations that shape individuals' life journeys.[9] Here, home embodies "a more abstract entity of belonging expressed through various types of narratives and other forms of symbolic interchange" (Fog Olwig 2002, p. 216). No matter how scholars interpret its nature, the home in its metaphysical form always denotes some positive meanings that motivate individuals to pursue a better life. While the economic values of a house implicate individuals in a relentless race for wealth and fame, the cultural values of home can nevertheless offer leverage for radical critique upon such a race (Young 2005). House ownership is currently a class-based privilege in a mass consumerist society, while a feeling at home is a subjective struggle for spatial and temporal autonomy. Understanding this difference helps to "de-myth the association of home with the domestic house" (McCarthy 2018, p. 980).

Another popular misconception is to treat home as equal to family life. In his influential book, *Haven in a Heartless World*, Lasch (1977) conceptualized the family as a refuge for the woman to support her exhausted husband and repair the spiritual damage inflicted by the market in nineteenth century Europe. When this traditional structure of working husbands and stay-at-home housewives collapsed, the family is besieged by the logic of market, and the sanctity of the home becomes "a sham in a world dominated by giant corporations and by the apparatus of mass promotion" (Lasch 1977, p. xxiii). The idea of home as a refuge, a fortress, or a haven in a heartless world reinforces the boundary between public and private, men and women, although this boundary is virtually blurred in the contemporary world.

In a different pioneering work on the idea of home, Douglas (1991) specifies it as a kind of space in which the mother operates the system of a public good for a family. This conception posits that a whole family, including parents and children, is essential to home and their daily activities are structured by mutual consultation and adjustment for functional requirements, for example, meals, sleep, relaxation. Once synthesizing home with family life, Douglas (1991) articulates tyrannies of the home over times, tastes, and even speech, which explains why so many people want to escape from the home. Additionally, the public good made by the mother is noncommercial. Scholars emphasize that the home is not a hotel, simply because fair access to resources at home should not be commercialized. As Douglas (1991, p. 300) explains, "the idea of the hotel is the standard 'Other,' where every comfort has to be paid for, the mercenary, cold, luxurious counterpart against which the home is being measured." If the home involves exchange, it is what Mauss (1990) calls a gift economy among family members instead of a money economy among unfamiliar persons. Duyvendak (2011) describes that in the United States, the stronghold of home draws on the dwelling of the nuclear family, long regarded as a haven in a turbulent world. In such a haven, men work outside, and women maintain home for the whole family. The US crisis of home, according to Duyvendak (2011), is deduced by the fact that women as the custodians of the traditional home have left the domestic sphere to enter the paid workforce as much as men.

Scholarship on the idea of home as family life has proven successful in exploring how home does not equate to a shelter or a house, but involves actual daily activities. Yet, this idea reveals some weakness. Feminist critics have made substantial contribution to complicate the nature of home beyond family life. More than 100 years ago, pioneering feminist scholar Charlotte Perkins Gilman (1903, p. 88) noticed that the operation of domestic life in a house drew on the economic dependence and specialization of women, since the daily processes of housekeeping are "a constant source of annoyance, friction, waste, and loss." The goal, generated by men, of living comfortably in a house also generated much discomfort and pain to women. Once their continuous efforts end with utter exhaustion, as Gilman (1903) suggested, women must go away from home for a rest, and the liberation of women requires getting them out of the house. Gilman's work has had much influence on feminist critics on the nature of home. For instance, Massey (2007, pp. 237–238) argues,

> The view of the home as the family's private retreat was closely linked to the new definition of woman's separate sphere, which glorified the role of the wife as a homemaker and full-time mother. In American society, the cult of domesticity that characterized this transformation in women's roles placed women on a pedestal as the custodians of the home and segregated them in their domestic sphere, while the public sphere was allotted exclusively to men.

Since women play an essential role in organizing family life, their feeling at home has been replaced by their supportive function for the good of others. For the sake of liberation, individuals need to get free of the constraints embodied in house to rebuild the temporal and spatial connection between themselves and the world. Following Gilman, Massey, and other feminist critics, I recognize the calls for a new understanding of home beyond materiality (house) and domesticity (family), although home is related to both.

Beyond house and family, we can discuss the making and unmaking of home. A critical geography of home developed by Blunt and Dowling (2006) constitutes a valuable explanatory framework that spells out the relationality of home: the material form of home depends on what home is imagined and experienced to be, and the imaginations of home are shaped by the material forms of dwelling. Furthermore, homemaking is an ongoing process of becoming, "a process creating and understanding forms of dwelling and belonging" (Blunt and Dowling 2006, p. 23). Homemaking is ongoing because it occurs over time and space as individuals negotiate the dynamic relations between symbols and materials, the self and the world, economy and culture, and it is political because this process is imbued with conflicts, confrontations, and compromises among various groups of people. As Blunt and Dowling (2006, p. 23) remind us:

> Home does not simply exist, but is made. Home is a process of creating and understanding forms of dwelling and belonging. This process has both material and imaginative elements. Thus people create home through social and emotional relationships. Home is also materially created—new structures formed, objects used and placed.

This statement indicates that homemaking is spatially situated in various home-geographies – apartments, communities, cities, and even the Internet – and in dynamic relationship with social networks to create various lived experiences. Furthermore, Blunt and Dowling (2006) argue for the need to exemplify the simultaneity of belonging and alienation in order to cultivate a politicized understanding of home. The making of home as material, imaginative, and lived inevitably incentivizes geographers to investigate the discrepancy that has resulted from the negotiation between individuals and the world.

Drawing on Blunt and Dowling (2006), Brickell (2012) argues that critical geographies of home are political by nature to reflect social and spatial inequality inflicted upon people in the margins. As one idealized site of human existence, the home offers a critical entry point to understanding and unraveling the micro-geographies of sociospatial uncertainty in a rapidly changing world. This uncertainty exemplifies the influence of wider structural forces of capital accumulation and political control upon our daily life. Thus, Brickell (2012, p. 227) notes that the home has been constructed by feminist researchers and other critical scholars as a radical political space in itself, but it is "a vulnerable crucible to wider political [and economic] processes." Furthermore, Brickell (2012) emphasizes that critical geographers of home must challenge hegemonic discourses of home and narrate and even participate in sociospatial acts of resistance that hold the potential to generate positive change in people's home(less) lives. With a focus on domestic life, Brickell's (2012) work is helpful in unravelling how a critical geography of home can illuminate and address domestic injustice. Similarly, Liu (2021) confines the time-spaces of home to the domestic sphere, arguing the need to explore domestic practices in time and space for homemaking. I recognize the value of mundane practices in the domestic sphere, but argue that this focus might dismiss the complexity of home. Rather I seek to analyze a broad and dynamic interaction between the self and the world at multiple scales.

Recent interest in home-city biographies aims to highlight the intertwined temporalities and spatialities of home and the city in people's everyday lives (Blunt, Ebbensgaard, and Sheringham 2021). For Blunt et al. (2021), urban homemaking is deeply shaped by housing precarity that creates structural contingencies of tenure, property ownership, and displacement. They also identify the value of home-city geographies to unravel the prevalence of unhomely life among those who cultivate a sense of rootedness in a turbulent housing market and who may be physically rooted yet feel a strong sense of affective displacement in the city. Such a geography of home-city complicates notions of alienation, loss, and exclusion, and pushes us to critically examine the dynamic relationship between home and homelessness. The entwinement between rootedness and displacement goes beyond the domestic sphere to project upon the urban scale, showing that urban residents feel both at home and not at home in the city. I recognize critical attributes and agendas of home in geography, and attempt to analyze the (un)making of home to exemplify the structural inequality – class, gender, ethnicity, and age – embedded in the process of China's great transformation. This inequality is particularly explicit among people who engage with mobilities.

Under conditions of extensive mobility, the making of home becomes more dynamic and disruptive. Instead of fixing home in space and time, a traveling culture reflects the fluidity of contemporary life and unsettles binary oppositions between tradition and modernity, here and there, past and present (Clifford 1997). Home is made "in a routine of practices, a repetition of habitual interactions, in memories, myths and stories carried around in their heads" (Nowicka 2007, p. 72). These propositions demonstrate that people on the move tend to release the stress of living in a permanent condition of homelessness, a condition coming out of dwelling on fast trains that run toward an unknown destination (Heller 1999). As Appadurai and Breckenridge (1988, p. 5) state:

> The world of the late 20th century is increasingly a cosmopolitan world. More people are widely travelled, are catholic in their tastes, are more inclusive in the range of cuisines they consume, are attentive to global media-covered events and are influenced by universal trends in fashion.

This cosmopolitan and traveling world has become even more central in recent years than it appeared. As mobilities entail a change of home or pursuit of multiple homes, they challenge the conventional understanding of home as a stable, fixed site that embeds individuals in a specific place, or that foregrounds a particular sense of identity.

Geographers have examined how home becomes a contested space for identity building and capital accumulation, in the cases of (im)migrants, diaspora, and refugees (Sheffer 2003; Blunt 2007; Blunt and Bonnerjee 2013; Adelman 2016; Boccagni 2017; Fathi 2021; van Liempt and Staring 2021). Low-end transnational migrants must make a new home to handle fear, danger, and the unknown in a foreign setting. To these migrants, turning away from the public landscapes of fear and danger involves "a turning toward home" where they can feel comfortable and secured (Ahmed 2004, p. 130). Life journeys in diaspora result from several interrelated cultures and geographies to navigate the desire to and myth of return. For those who actually return, the homeland perhaps remains static, but these returnees have a different horizon after a long journey elsewhere. The look toward homeland is a look "back" into a past moment, such as childhood or youth, or a society's historical past. This look can become uncanny simply because "our past, whatever it was, was a past in the process of disintegration; we yearn to grasp it, but is baseless and elusive; we look back for something solid to lean on, only to find ourselves embracing ghosts" (Berman 1982, p. 333). Unlike refugees who desire to settle down permanently and put down roots so that they can carve out spaces of belonging in a new homeland, expatriates enact a form of flexible attachment that reflects a dynamic and changing social and spatial relation between themselves and the places they dwell (Cai and Su 2021; Kunz 2016). To expats, the ideal of home is not confined with citizenship and housing, but rather associated with "a yearning for connectedness, or social relationships that enable them to feel validated as individuals" (Wiles 2008, p. 128). The experience of homemaking by mobile people involves an endless process of uprooting and regrounding so that they can negotiate the meanings of belonging and attachment (Ahmed et al. 2003).

Regardless of their social and economic background, mobile individuals face changing relationships between the self and the home. Geographers who critically study home emphasize that people on the move build their home by making and reconfiguring the destination both spatially and culturally, exemplifying a sense of home on thresholds "between memory and nostalgia for the past, everyday life in the present, and future dreams and fears" (Blunt and Varley 2004, p. 3). Aware of this condition, Ahmed (1999, p. 78) rejects the idea that home-dwelling and away-traveling are oppositional experiences, arguing that home is not "a place with boundaries that are fixed," but rather encompasses both familiarity and estrangement. Within the condition of being-in-the-world and becoming-in-process, homemaking is spatially situated in various home-geographies – apartments, hotels, communities, and cities – and in dynamic relationship with social networks to create different lived experiences. In essence, *homemaking* means an endless journey to reconcile the desire for homely life in a relentless and restless world. This journey reflects an exercise of human agency by individuals who make choices about their time and space within the constraints and opportunities brought by the external world.

With homemaking, time matters. As Tuan (1980, p. 6) specifies, rootedness implies "being at home in an unselfconscious way." I call it *rooted home*. Meanwhile, a sense of home in a place implies "a certain distance between self and place which allows the self to appreciate a place" and develop some positive feelings toward it (Tuan 1980, p. 4). I call it *sensed home*. While rooted home results from familiarity through long residence, particularly during childhood, sensed home draws on "a knowing that is the result of conscious effort" in a relatively short period of time (Tuan 1980, p. 8). For most people in the modern world, their home lies somewhere in the middle range of these two modes. Here, the rooted home, crystalized in past and familiar landscapes, is the home left behind, yet exists in dream or memory, with a hope to return one day. Meanwhile, the sensed home is constantly remade and recreated in various localities to achieve a temporary moment of inner calm in a relentless world. Due to social acceleration and high fluidity in an era of globalization, the possibility of making and maintaining rooted home diminishes, and the need to make sensed home(s) "with familiarity and ease, with the assurance of nurture and security" becomes necessary to counter the encroachment of homeless mind (Tuan 1977, p. 159).

As a research subject on its own, home embodies cultural significance and subjective feelings. From a humanistic perspective, renowned geographer Yi-Fu Tuan (1977, p. 149) proclaims that "home is at the center of an astronomically determined spatial system. A vertical axis, linking heaven to the underworld, passes through it." Similarly, Bachelard (1964, p. 72) argues that the home is "a key element in the development of people's sense of themselves as belonging to a place." Such a central position can inspire imaginations and feelings that motivate activities for a relatively better life. Indeed, home, either rooted or sensed, entails familiarity and attachment and touches upon the kernel of humanity, but experiencing such a home requires complex transactions to negotiate the deal of modernity. The time has come for research in the literature of home to engage more squarely with mobilities, feelings, experiences, and the exchange between meaning and money. This is not just a matter of what the home is, but such inquiry points to some critical agendas, such as how individuals imagine and experience home and for what purposes. Furthermore, home carries and embodies a great diversity of cultural and political understandings which differ among social groups and within geographical contexts. I now shift to a sense of home in traditional Chinese culture.

A Sense of Home in Traditional Chinese Culture

A Chinese conception of home can be traced back to inscriptions on oracle bones of the Shang Dynasty (c. sixteenth to eleventh centuries BCE) depicting roofs above pigs. A home (*jia* 家) in this context meant a concrete structure where pigs were raised. This conception gradually expanded to encompass the sense of comfort and emotional attachment that can be found in a dwelling place for humans. Beyond house (concrete building) and domestic life, home in the Chinese context is often associated with rootedness or place of origin. Home is both an adjective and a noun. As an adjective, it describes a land (jiaxiang, 家乡), a family (jiating, 家庭), or a country (jiaguo, 家国). As a noun, it generally links with "go" as in "go home" (huijia, 回家). This section delves into traditional Chinese culture to outline how a sense of home has been articulated. The goal is not to romanticize a traditional society or yearn for the loss of these cultural elements in contemporary China. Rather, I bring attention to how these elements influence people's lives then and now.

To start, I want to briefly clarify one key philosophical idea in Chinese traditional culture. The unity between human and nature (tianren heyi, 天人合一) represents a basic principle of existence, which means that the Chinese rarely place themselves above nature.[10] In China's cultural tradition, the doctrines of different philosophical strands, including Taoism, Confucianism, and Buddhism, highlight the value of the unity of humankind and nature as a foundation that underpins the relationship between the self and the world. In ancient China, the term tiandi (天地, heaven and earth) is a metaphor of the natural environment. While earth enables individuals to live in abundance, heaven brings sunlight and rain, in line with cyclic temporalities (seasons and the cycle of sunset and sunrise). For instance, Zhuangzi proclaims that

> 天地者，万物之父母也，Heaven and Earth are the father and mother of all things.
> 合则成体，It is by their union that the body is formed;
> 散则成始 it is by their separation that a (new) beginning is brought about.[11]

Nature has its own ways of operation and nourishes humanity. Out of this unity lies a conception of home in traditional Chinese culture as a destination for return and a balanced way of living in the world. Before elaborating this conception, I emphasize that these doctrines in Chinese cultural traditions enshrined an explicit class and gender dimension because only elite male intelligentsia had the privilege to leave home and return, and to pursue self-cultivation (xiushen, 修身), and to enjoy idyllic environments for dwelling and relaxation.

Home as a Destination for Return

As the renowned Chinese scholar Qian Zhongshu (1997) has pointed out, China's ancient thinkers, particularly Daoists and Zen Buddhists, always used "going home" as a metaphor to describe the journey for spiritual ending.[12] According to Qian (1997, p. 542), this metaphor exemplifies an endless exploration to discover the self and search for the meanings of being in the world:

> "Going" means a route and "home" the objective of such a route. Route includes many rounds of reversion and departure to seek peace; home refers to a familiar, congenial, and accustomed place where people can rest their soul calmly. (Author's translation)

Going home is a discovery of self and a search for a spiritual ending that fulfills the course of one's life. This spatial imaginary, which is prevalent in Chinese cultural traditions, reflects the idea that home is a representation of rootedness, made in and through different kinds of journeys stretching far beyond the presence of a physical dwelling. Going home becomes a journey to cultivate such a unity between human and environment so that elite intellectuals can discover the ultimate meanings of life.

Where is the destination? In China's rich traditions, two kinds of destinations stand out as the most meaningful symbols of home. The first one is internal, meaning human nature, which refers to the internal world of the mind and contemplation. Heavily influenced by Taoism, Bai Juyi (白居易), one of the greatest poets in the Tang Dynasty (618–907), poetically noted that home is where he can feel at ease:

> 我生本无乡 I am born without a homeland
> 心安是归处 the place where my heart calms down is a point of return (Author's translation)[13]

This understanding of home highlights the importance of inner calm and contributes to the key idea of self-cultivation in Chinese cultural traditions. By reflecting on their own actions and words, individuals can pursue the peace and purity of their own mind. Going home means a journey to reach deep into their own mind for the sake of spiritual satisfaction beyond material gain and scholarly fame. As Ma and Liu (2017, p. 13) assert, traditional elite intellectuals in China regarded the inner harmony as "the basis for harmony among all things in the world." In this sense, a return to inner nature can help to maintain an order for human life.

The second destination is external nature, the natural environment in which human beings dwell. Being at home means embedding one's body and mind in an idyllic natural environment imbued with mountains and rivers. China's tradition of water-and-mountain (shanshui, 山水) or pastoral landscape (tianyuan, 田园) poetry, which originated in the early fifth century, represents the earliest engagement with wilderness and natural environment, an engagement highlighting China's traditional elite intellectuals' embrace of the "natural world" as part of their spiritual cultivation (Hinton 2005). By articulating a deep sense of belonging to waters and mountains, this cultural tradition spells out the pursuit of the unity of humankind and nature as a dynamic cosmology through which individuals can feel at home in the most fundamental way. Representative figures in this tradition included Tao Yuanming (陶渊明, 365–427), Xie Lingyun (谢灵

运, 385–433), and Wang Wei (王维, 699–761). To these great scholars, the spiritual value embodied in the natural world offered ample solace to enrich their life.

The love for natural environment is best expressed by Tao Yuanming. In accordance with Taoist philosophy, Tao chose to stay away from officialdom and retreated in idyll. In his famous poem *Returning to the Farm to Dwell I*, Tao described his joy of going home after long years of service in the administration. At the beginning of this poem, Tao specifies that

少无适俗韵，性本爱丘山。	Nothing like all the others, even as a child, rooted in such love for hills and mountains.
误落尘网中，一去三十年。	I stumbled into their net of dust, that one departure a blunder lasting thirty years.
羁鸟恋旧林，池鱼思故渊。	But a tethered bird longs for its old forest, and a pond fish its deep waters.
开荒南野际，守拙归园田。	So now, my southern outlands cleared, I nurture simplicity among these fields and gardens, home again.[14]

The verses demonstrate that Tao gave up the pursuit of professional ambition and scholarly fame and returned to home located in a natural environment. Tao used the metaphors of birds and fishes to describe his longing for home, that is, closeness to the natural world so that he can discover his true self. Tao's choice cannot be simply explained as escapism or sentimental pastoralism, but rather a return to home, the ultimate destination of his life journey ending with the unity of human and nature. At the end of this poem, Tao reiterated that "Back again: after so long caged in that trap, I've returned to occurrence coming of itself."[15] These two verses further exemplify that a return to the natural environment in quietude and simplicity allows Tao to return to the origin of his life.[16] This return or escape represented a meaningful form of reclusion by Chinese traditional literati to synthesize external nature and internal nature into one: the home.

To experience a reclusive culture, Tao Yuanmin chose to live a simple life and adhere to spatial and temporal autonomy. Without monetary support from the imperial court, Tao and his family had to live frugally and sometimes borrowed rice from local farmers (Hinton 2005). Unsurprisingly, Tao has been regarded as a legendary figure who almost realized and practiced the ideal of reclusion. To China's cultural elites schooled in Confucian doctrines in the past, aspirations to serve state and society remained a strong commitment. Nevertheless, some chose not to follow this path, but engage in reclusion as an attractive alternative to the obligations to public duty.

Reclusion into a natural environment, as Tao chose, has always lingered in the minds of Chinese cultural elites in the literati class's pursuit of inner calm and self-cultivation. As Swartz (2008) explains, the idyllic side of Tao's reclusion and its many pleasures, including wine, music, books, poetry writing, a sense of freedom, and a leisurely life, has provoked resonance among Chinese people ever since the Six Dynasties (222–589), although Tao's efforts to handle material deprivation and daily hardship were neglected. Cultural critic Gao Dapeng highlights Tao as a unique figure to reveal reclusion (yinju, 隐居)as a lingering longing among Chinese traditional literati.

> Whether in terms of nature, society, culture, or religion, the term *yinju*, 隐居 [reclusion] carries its own historical tradition; so we can say that it had become the "collective unconscious" of the Chinese literati. It had become "internalized" as part of the intellectual's personality … it was the most fundamental, most pervasive, and even highest value. … Yin was a moral ideal for the Chinese literati, as well as an aesthetic and an overarching ideal. In the long-term progress of history, it had become the common dream of the literati. Translated by and cited from Swartz (2008, p. 11)

Such a reclusive culture carved out an alternative lifestyle beyond seeking higher positions in official public service, the most popular mission pursued by talented and learned elites in ancient China (Tian 2005). Together, Tao and others in the same period created a demeanor to enrich the spiritual world of intelligent people in China. The longing for reclusion in the past evolves into the contemporary yearning for temporary retreat in idyllic destinations where it is possible to enjoy some pleasures that Tao Yuanming displayed to his readers over 1,500 years ago.

A sense of home in ancient China does not depend on bricks or mortar, as it can be located in a mountain, a forest, or a piece of land, or simply in the heart. Indeed, internal nature and external nature are intertwined to cultivate a feeling of being at home in this actual world. Generally, closeness to the natural environment enables individuals to calm down so that individuals can poetically enjoy natural landscapes. In a response to the question: where is home, Li Bai (李白, 701–762), one of China's greatest poets, specified that "I make my home in the mountains." He explains in a poem,

问余何意栖碧山,	You ask why I live alone in the mountain forest,
笑而不答心自闲	and I smile and am silent until even my soul grows quiet:
桃花流水苜然去	The peach trees blossom and the water continues to flow
别有天地非人间	it lives in the other world, one that no one owns.[17]

This poem is a typical example of enjoying the natural environment for the sake of inner calm. Beautiful landscapes, peach trees, running rivers, and so on contribute to a sense of home in line with wilderness and naturalness. For sure, the highest ideal of Chinese thought is that a person does not need to escape from human society in order to preserve his or her original, happy nature, but becomes a second-rate recluse to maintain a condition of solitude (Lin 1998). A trip to, and time in, a natural environment can definitely bolster such a sense of home and facilitate self-cultivation.

Home as a destination of return entails a deep cultural meaning to Chinese people at large. As Kuhn (2008) observes, Chinese overseas emigrants to Southeast Asia and other lands since the Qing Dynasty rarely thought of themselves as permanent settlers, which made them different to European explorers who wanted to conquer and colonialize foreign lands. Rather, Chinese emigrants maintained monetary and cultural connections with their homeland and dreamed that one day they would return for eventual settlement in the land where they were born. Due to this nostalgia for home, Kuhn (2008) defined this spatial relocation as a form of sojourning, which means living temporarily away from home with the intention of returning sooner or later. This return became common in Chinese life and very few could resist the call of the home, giving rise to the stay-at-home myth. The myth means that the sojourners needed to maintain ties with folks from the same hometown and then form regional associations (tongxiang hui, 同乡会) so as to survive and thrive in alien places. Another key job done by regional associations was to settle posthumous affairs. Once a sojourner passed away in an alien land, it was believed that the deceased became a wanderer forever and degenerated into a lonely ghost. To settle the sojourner's affairs, the regional association would bury the remains in a compatriots' cemetery or even send the remains back home for burial so that kinsmen could "perform funeral rites for the repose of his soul" (Kuhn 2008, p. 44).

Another example of going home comes from the prominent humanist geographer Yi-Fu Tuan. Tuan (2007) aptly captures how his return to China becomes a journey of going home after he had been absent for more than 60 years. He explicitly expresses his feeling: "I was a tourist in China. I even went to the Great Wall! But I was also a native who had come home" (Tuan 2007, p. 175). This juxtaposition between home and traveling raises the question of identity and belonging. This question becomes more discernible when Tuan is disturbed about how to describe where he belongs and whether he is a Chinese, an American-Chinese, a Chinese-American, or an American. Being born in China, Tuan maintains his ties with Chinese culture by reading Chinese history and geography books when he lived in the United States. Despite this long departure from China, Tuan's going-home journey has opened his eyes to "the possibility of happiness" that, according to him, requires "grounding in total acceptance, intimate ties, and a love that does not need to be repeatedly won" (Tuan 2007, p. 174). Hence, his journey to China illustrates what a Chinese sense of home is like. This imaginative projection of home explains the representation of home in Chinese culture as a destination of return after a life journey, although this journey is open-ended and unpredictable.

Today, the term "going home" exemplifies a journey of visiting the hometown for a reunion with family members, particularly elderly parents. To hundreds of millions of residents living in cities, their parents are often cited as a destination of return for cultural roots. While Qian Zhongshu's poetic illustration of going home specifies a homely place where people can rest their soul, an existential expression of going home means visiting elderly parents who live in the hometown left behind. While paying a visit to elderly parents indicates the fulfillment of filial piety, it also represents a cultural journey to renew a sense of belonging, a sense rooted in the social and blood ties with parents. Going home has been internalized by the Chinese as a journey of return, although where to return to is subject to different interpretations. Such a journey influences their individual temperaments in relation to the surrounding. Between one journey and another lies a balanced way of living at home.

Home as a Balanced Way of Living

Home as a way of living means that people live within an earth-bound world rather than a spiritually abstract utopia. In his classic work *The Importance of Living*, which was originally published in 1937, Lin Yutang (Lin 1998, p. 23) emphasizes that "all questions of living in heaven must be brushed aside. Let not the spirit take wings and soar to the abode of the gods and forget the earth." The reality and spirituality of life, which is located here and now, indicates that ancient Chinese philosophy is preoccupied with the knowledge of living.[18] Among many questions, the eternal one is: how are we to live in this mundane world? More specifically, how shall individuals organize their life so that they can live comfortably in this world? Indeed, ancient Chinese philosophers rarely considered how to achieve an immortal life in a heaven, but rather addressed the way or art of living here and now.

The way of living means how to live, not in a stationary manner, but with constant change. The term "way" itself encompasses pathway or roadway, that is, the process through which all things arise and disappear (Hinton 2005). At the very beginning of *Dao De Jing* (道德经), Laozi specifies that "the Dao [or way] that can be trodden is not the enduring and unchanging Dao."[19] The seasonal cycle is one manifestation of the Dao to naturally organize and orchestrate things in the world. Following cyclic time is a natural or spontaneous way of living, and any effort to forcefully interrupt the operation of cyclic time can disrupt the natural process of change and thus generate negative consequences. The way of living does not necessarily mean that individuals must become a recluse from the mundane world and retreat to a remote mountain. Rather, the key is to follow cyclic time to enjoy a balanced lifestyle between busyness and loafing.

Ancient philosophers in China pointed to the need to avoid overwhelming busyness. Otherwise, individuals would have no time to enjoy themselves, develop meaningful reflection for inner peace, or rest their mind. The essence of a good balance has been aptly narrated in Chinese traditional dictums, including the Confucian principle of golden mean, Laozi's call for letting things take their course, and the Buddhist value of the state of self-oblivion. As Confucius commented on how ordinary farmers enjoyed carnivals after a busy season of work in the field,

> For their hundred days' labour in the field, (the husbandmen) receive this one day's enjoyment (from the state); this is what you do not understand. (Even) Wen and Wu could not keep a bow (in good condition), if it were always drawn and never relaxed; nor did they leave it always relaxed and never drawn. To keep it now strung and now unstrung was the way of Wen and Wu.[20]

This dictum indicates that busyness interspersed with loafing is the way in which King Wen and King Wu do things in a balanced way. Such a balance between these two typical tempos – busyness and loafing – generates a natural cycle of time to reflect the principle of unity between human and nature. Being aware of the danger of overload, Confucius kept emphasizing the need to slow down and even retreat into a natural environment so that each one can conduct self-reflection and reach into their inherent nature: "It belongs to the nature of man, as from Heaven, to be still at his birth. His activity shows itself as he is acted on by external things, and develops the desires incident to his nature."[21] Hence, home as a destination of return and an experiential way of living are united into one category of mundane life.

Confucius's call for the way of Wen and Wu was shared by other scholars. For instance, Zhuangzi emphasized the need to calm down for inner peace and to conduct daily activities in accord with cyclic time. What he called a happy excursion (xiaoyao you, 逍遥游) expresses how to appreciate the great beauty of heaven and earth. The art of living in the world, as taught by Tsesse, Confucius' grandson and author of *The Golden Mean*, is the way of sweet reasonableness (Zhongyong zhidao, 中庸之道), that is, "a perfect balance between action and inaction, shown in the ideal of a man living in half-fame and semi-obscurity; half-lazily active and half-actively lazy" (Lin 1998, p. 113). The art of living is essentially the art of loafing, and moreover, the wisest persons are those who loaf most gracefully. As Lin (1998, pp. 150–151) explains in detail, "the philosophy which nourishes this divine desire for loafing in China ... gives rise to that carefree, idle, happy-go-lucky—and often poetic—temperament in the Chinese scholars, and to a lesser extent, in the Chinese people in general."

The significance of loafing becomes particularly explicit once everyday life has been accelerated and disciplined in modern times. Regarding the term loafing, I am cognizant of the negative connotation related to it and other similar words such as idleness or laziness. For instance, Oxford Advanced Learner's Dictionary defines "loafer" as "a person who wastes their time rather than working." Although I could choose some banal words such as relaxation or leisure to describe a condition in which individuals spend time for un-utilitarian purposes, I realize that these words are constantly mixed with commercial intent and related activities have become an important part of the tourism industry in a broad sense.

In this book I nevertheless intentionally choose loafing to describe a condition of relaxation and comfort without being harassed by time discipline. Despite that the term has negative connotations in the context of capitalism, loafing holds conceptual value and reflects a cultural undertaking in traditional Chinese culture. The meanings inherent in the concept of loafing may give rise to associations of temporal autonomy through active choices.

How does this framework of loafing help us to understand homemaking in post-Mao China? One thing is certain. Busyness has overwhelmed loafing as the key to organize daily life. It is time to bring loafing back as a key element for a balanced way of life. I conceptualize loafing as a subjective experience of spatio-temporal autonomy through which individuals conduct unproductive activities. It contradicts the dominant temporality of modern societies characterized by speed, efficiency, and functionality. It is similar to leisure that, according to Veblen (2007[1899], p. 33), does not connote indolence or quiescence, but "non-productive consumption of time." Nonproductivity is a key feature of both idleness and leisure. Veblen (2007[1899], p. 33) emphasizes: "Time is consumed non-productively (1) from a sense of the unworthiness of productive work, and (2) as an evidence of pecuniary ability to afford a life of idleness." Today, leisure has constantly been filled with purpose-driven obligations, shaped by economics (profit in the leisure industry) and politics (social prestige and class difference). Thus, Hemingway (1996, pp. 27–28) asserts that leisure has deformed through "increasing commodification and consumerization," reflecting and reinforcing a growing instrumentalism in leisure time. Once the original sense of freedom in leisure has given way to consumerism, leisure becomes associated with "a materialistic style of life and a means of control" (Juniu 2000, pp. 71–72).

Loafing is different from leisure. Instead of a performance of social status, loafing highlights subjective and reflective experiences of spending time free of utilitarian purpose. If there is a purpose, it aims for self-cultivation and inner calm. These experiences are characterized by Lefebvre (2004) as banal, subtle, and spontaneous, in harmony with cyclic time. In this sense, loafing likens to a spiritual retreat and denotes the practice or state of *otium*, that is, "repose, rest, inner calm or, quite simply, a liberation from duties and responsibilities, from schedules and expectations" (Fludernik and Nandi 2014, p. 2). In contrast with time discipline, loafing means that individuals reject the obligation to clock watch and embrace a condition of inactivity and nonproductivity (Crary 2013). It is not about doing typical things with real purpose, as too much doing is not conducive to calming down, but about redoing and even undoing banal activities.

No matter how much Lin Yutang highlights the value of loafing to a homely way of life, the experience of loafing has degenerated into an endless process of demonization in the era of modern capitalism. To put it simply, time is regarded as a core structural component of capital accumulation and thus cannot be wasted on nonproductive activities. If ancient Chinese philosophers were right, as they reiterated the need to strike a balance between busyness and slowness and promoted a loafing lifestyle, speed and the acceleration of life inevitably perpetrate the loss of home feelings and deduce a crisis of home. In the mid-1930s, Lin (1998, p. 148) lived in New York and observed that "the American is known as a great hustler, as the Chinese is known as a great loafer." The former was caused by the implementation of machine civilization decreed by technological advancement and time discipline, while the latter was ingrained in a traditional way of living without succumbing to modernization. Lin (1998, p. 162) went further to summarize that there were three great US vices: efficiency, punctuality, and the desire for achievement and success. This observation is echoed in the literature on a loss of home in the United States. For instance, Duyvendak (2011) summarizes scholarly works to succinctly state that US citizens suffer from time discipline, feel overworked, and have little time for home. In a poetical tone, Lin (1998, p. 162) mentioned that time discipline deprived US citizens of "many a good, idle and beautiful afternoon." One of Lin's prophetic visions was that Manhattan would slow down and the US hustler would become an Oriental loafer. Learning to be a loafer might be a palliate to time discipline. Having said this, Lin Yutang did not have a chance to predict the exercise of rational efficiency in today's China. Nor did he live long enough to watch China's addiction to speed, an addiction even worse than that in the United States in the late 1930s.

The wealth of insights provided in traditional Chinese culture on the cultural construction of home holds the potential to be more fully integrated into emerging analyses of unhomely life in contemporary China. The home represents a destination of return, either to external nature or internal nature, to rest the soul and achieve inner calm. Furthermore, the home embodies a balanced way of life between busyness and loafing in this actual world, and homemaking holds the potential to fulfill humanity. Alongside the celebrations of busyness and speed as a triumph of modernization over human capacities, loafing can emerge as an alternative tempo to bring back the social meaning of time. After all, Chinese cultural traditions emphasize the value of loafing for the purpose of self-cultivation rather than self-enslavement.

So, what make a Chinese notion of home unique? First, it is deeply embedded in soil-based rurality. The farming culture in Chinese society has existed for thousands of years and true love for the soil has been deeply ingrained in the mentality and imagination of Chinese people. Fei Xiaotong's work on *xiangtu* China illuminates the rural root of Chinese culture and explains why Chinese people have such a strong passion for the soil. Second, a Chinese home represents a secular life in this actual world. It is neither a heaven nor a haven, where an afterlife unfolds to allow individuals to enjoy happiness and comfort.[22] Both ancient elite intelligentsia such as Tao Yuanming and today's urban middle classes struggle between imposed cage and spatiotemporal autonomy in their everyday life and work hard to maintain moments of loafing. Finally, home is symbolized as a destination – external natural environment and inner freedom – towards which going becomes a lifelong journey of self-cultivation.

After articulating the uniqueness of a Chinese notion of home, I must be careful about two aspects. First, this articulation inevitably simplifies the cultural notion of home in China, especially since I intend to delve into the essence of Chinese ancient philosophy. Instead of a meticulous interpretation of what is home in China, I intentionally seek inspiration from Tao Yuanming, Lin Yutang, Fei Xiaotong, and Qian Zhongshu to highlight the spatiality, temporality, and spirituality of home in the context of China. Indeed, the cultural tradition of home in China has been complex and full of interpretations and conflicts, and the meanings of home have been constantly changing and evolving. Second, I try to avoid a framework of China versus Europe. As Tian (2005) warns, any discussion of the uniqueness of the Chinese experience results from one's ignorance of other cultures and civilizations. Here, I draw on farming culture, Confucian tradition, and the literati spirit to conceptualize the culture of home in China. My interpretation of home in China aims not to set one against the other, although I cannot fully avoid reductionism.

Indeed, a Chinese notion of home shares much common ground with ideas of home in other geographic contexts. Home entails some common values for every human being in the world, including safety, a sense of security and control, ties and bonds, and autonomy (Wiles 2008). Writing about the poetics of space, Bachelard (1964, p. 4) asserts that "our home is our corner of the world … it is our first universe, a real cosmos in every sense of the word." Regardless of what constitute *the* home, we can argue that home can be universally conceptualized as a center of human existence with safety and security and an ensemble of positive values projected by individuals upon their surroundings. Going further, Young (1997, p. 162) argues that at home individuals have the ability to preserve "the meaningful things in which one sees the stories of one's self embodied, and rituals of remembrance that reiterate those stories." To Young, making and preserving meanings remain essential to home, for both individual and collective identities. The association between meaning and home is a key to understanding the culture of home in China, too.

Modernization and Loss of Home Feelings in Post-Mao China

In *The Consequences of Modernity*, Giddens (1990) notes that the profit-driven order comes about in social systems to bind time and space. Specifically, time and space are organized to connect presence and absence, giving rise to the separation of time and space and the disembedding of social systems (Giddens 1990). In a different manner, Harvey (1989, p. 240) contends that the conditions of modernity draw on the time-space compression that revolutionizes "the objective qualities of space and time" towards disorientation and alienation. Regardless of separation or compression, the meaning and use of time are changed in an era of modernity, starting with the Industrial Revolution. While time was used for tasks in peasant societies, it was spent for profit in industrial societies (Fei 1992). Accompanying this change was the deployment of linear time around the clock to rationalize production and facilitate labor exploitation.

The historian E. P. Thompson captures the implementation of the clock to institutionalize the use of time among factory workers in England. As Thompson (1967, p. 61) states succinctly, "time is not currency: it is not passed but spent." Indeed, temporal efficiency became a dominant rule to be accepted and internalized by factory workers in England.

> The first generation of factory workers were taught by their masters the importance of time; the second generation formed their short-time committees in the ten-hour movement; the third generation struck for overtime or time-and-a-half. They had accepted the categories of their employers and learned to fight back within them. They had learned their lesson, that time is money, only too well. (Thompson 1967, p. 86)

This dramatic change is also described in detail by Foucault in his classic work *Discipline and Punish*. At the dawn of European industrialization, one key task of the exercise of power was to prohibit laziness, or the existence of nonproductive time: "The body becomes a useful force only if it is both a productive body and a subjected body" (Foucault 1979, p. 26). The correct use of the body was built upon a correct use of time. To achieve the goal, disciplinary control was taught and imposed, generating a condition of what Foucault calls time discipline. Foucault (1979, p. 154) further elaborates:

> [I]t was forbidden to waste time, which was counted by God and paid for by men; the timetable was to eliminate the danger of wasting it—a moral offence and economic dishonesty. Discipline, on the other hand, arranges a positive economy; it poses the principle of a theoretically ever-growing use of time: exhaustion rather use.

Henceforth, the idle individual, who could not comply with the norms of production, had to be disciplined and punished. While education aimed to organize "a new economy of the time of apprenticeship," prisons functioned as a space of disciplinary order to punish those who wasted time, including beggars and idlers (Foucault 1979, p. 147). Time discipline has come to dominate the meaning and use of time well beyond the realm of production, and has penetrated deeply into everyday life.

This dominance occurred in China when the country started massive modernization after 1978. Once liberated from the Chinese state's meticulous control over their private lives during the Maoist regime (1949–1976), individuals in the post-Mao period eagerly engaged in economic reform. In late 1978, Deng Xiaoping came into full power after the Third Plenum of the 11th Communist

Party Congress. Immediately he embarked on his experiment of developing China's market economy by a policy of Reform and Open (*gaige kaifang*, 改革开放). The purpose was to enable China to "catch up" with the most developed countries in the world. To be rid of existing ideological impediments, Deng introduced a philosophy of economic pragmatism to replace the legacies of a planned economy and to remove the dogmatic ways in which people strictly followed the instructions of the great leader Mao Zedong. In an interview in October 1985 by Henry Grunwald, editor-in-chief of *Time*, Deng acknowledged that "the overriding task in China today is to throw ourselves heart and soul into the modernization drive."[23] One pragmatic approach to achieve modernization is to "let some people and some regions get rich first" (Deng 1993, p. 149).

In response to the call to get rich first, the market economy became the new social order to instruct people's everyday life. The Chinese state launched a nationwide campaign to break the shackles of old habits and urge people to liberate their thought. Here, to liberate meant to set free from the fetters that obstruct the race to get rich first. To earnestly launch the task of building a new order, first all those shackles, such as traditional loyalties and refined loafing, which once characterized China's *Xiangtu* society, had to be discarded. Economic reform meant first and foremost dismantling the unnecessary obligations standing in the road of market economy and liberating potential laborers from "the shackles of the family-household duties and from the dense tissue of ethical obligations" (Bauman 2000, p. 4). Economic reform, in its own way of reorganizing time and space, completes the tasks of building a new society that the Maoist revolution had been unable to bring to fruition. Indeed, the reform generates two profound consequences: the fall of the planned economy as an institutional logic of organizing everyday life, and the rise of market economy to engulf everyone in an endless game of getting rich first and fast.

Following Deng Xiaoping's tutelage of "get rich first [and fast]," the desire for wealth and fame took a firm hold in the 1980s. Accordingly, time became a medium for profit, attentive to Deng Xiaoping's phrase that "time is money," a popular slogan indicating that speed has become the core value of the grand project of modernization in post-Mao China. The promise of richness via hard-working ethos has been totalizing in its power to instruct and even force individuals to maximize the economic value of time in the process of production and consumption. By the end of the 1990s, the quest for materialistic gain was deeply rooted in China to such an extent that individuals accepted the cult of speed and internalized a teleology of busyness into their mentality. The relentless process of modernization has produced a society populated with self-disciplined and bustling individuals with an intent on monetary gain. The result, as Harvey (2010a, p. 83) proclaims, is that people are condemned to "live in a topsy-turvy world of material relations between people and social relations between things." The best example is the "9-9-6" schedule (working from 9:00 a.m. to 9:00 p.m., six days a week) in IT firms in China's big cities. Reporting on a nationwide debate in April 2019 on this high-pressure schedule, He Wei from *China Daily*, the largest state-owned English newspaper, noted that this industry malaise forces workers to conform to a relentless work culture.[24] Scholars have revealed that the forces of state, capital, and technocracy in Chinese cities have engulfed ordinary people's everyday life with endless commodification and alienation (Wu 2018; Zhang 2018).

The grinding 9-9-6 work culture is widely used in some of China's most prestigious IT corporations. To borrow the words of Jack Ma, the founder of Alibaba (one of China's largest IT corporations), this culture represents a fortune bestowed on those who work hard and earn high salaries. The condition in some corporations is even worse, due to a new work ethic coded as 007, which means working from midnight to midnight, seven days a week, and resting only on rotations. Both work cultures overtly defy the Labor Law enacted in 1995. Article 36 in the law stipulates that "The State shall practise a working hour system wherein labourers shall work for no more than eight hours a day and no more than 44 hours a week on the average," and Article 38, "The employer shall guarantee that its labourers have at least one day off a week."[25] Indeed, Article 41 ensures that the employer can prolong work hours due to needs of production or business, but overtime work shall be no longer than one hour a day. Ironically, 9-9-6 and 007 overtly violate the law, but no firms have been punished so far. Crary (2013, p. 9) uses 24/7 to describe a global temporal infrastructure for continuous work and consumption, which is built upon "a non-social model of machine performance and a suspension of living that does not disclose the human cost required to sustain its effectiveness." This suspension is particularly true in China. The immediate result is a loss of home feelings. In the following, I will explain three major factors accounting for this loss.

First, the meanings of time are lost in the process of modernization. The relentless process of modernization in post-Mao China, in which time discipline assumes a foremost logic of sociospatial organization, has deduced the acceleration of life. To get rich first and fast, people measure their time in terms of economic productivity that further reinforces time discipline in work, recreation, and other parts of their daily life. Thompson (1967, p. 95) observes, in the case of the British working class during the Industrial Revolution, that people were fascinated with "the desire to consume time purposively, which most people carry just as they carry a watch on their wrists." This desire similarly has happened in the post-Mao Chinese society in which individuals are enticed to efficiently chase social advancement.

The obsession with speed and productivity has had a dramatic effect on the everyday life of individuals in China, who have become mesmerized by a seemingly unlimited pursuit of wealth and fame. My respondents explain how busy they were:

> **Yali:** In Beijing, I don't have time to think about how to kill time. I was either working or falling asleep. I needed to work 14 to 16 hours every day and faced enormous occupational pressure. There is no such thing as relaxation.

Liu: People in big cities revolve around office, home, and school with a hasty pace during weekdays. On the weekend, they are not free because they have to take care of kids, parents, and unfinished work. They have so many things to care about.

These two quotes offer a glimpse of busyness in China's big cities. While Yali, who is a business owner of an Internet company in Beijing and visits Lijiang as a tourist, must devote his time and energy to full capacity, Liu describes a common rhythm that constitutes individuals' daily tempo. To stay away from speed, Yali chooses to visit Lijiang for a short break and Liu decides to migrate to Lijiang and keep a distance from such a hectic rhythm. Individuals like Yali and Liu in Chinese metropolis are living in a world obsessed with speed, with doing things fast and efficiently. Similar to Yali and Liu, many middle-class Chinese feel alienated by the world around them, a world that undertakes rapid modernization, and troubled by "confusion, emptiness, and uncertainty, despite living a materially comfortable life" (Zhang 2020, p. 137). Their daily life in metropolis is like a race against the clock, cramming more and more things into a day. They may not necessarily see the clock, but it orders their daily life practices and prunes nonproductive use of time, while simultaneously perpetuating the economization of time.

The obsession with speed and busyness is captured by other scholars. For instance, Mason (2020) provides in detail how individuals in public heath units in Guangdong province have to sacrifice their off-duty time to participate in work-related activities. To these professionals, there is no such a thing as the separation between business time and private time. As Mason (2020, p. 856) describes, they must "work late, go banqueting, or do any other time-consuming tasks that conflicted with their personal priorities." Regarding why they make the sacrifice, the common reply is that these professionals either build guanxi with colleagues or toe the line so that they can maintain a good job performance. This work-forever mode leaves them with little control over how they can spend their spare time. Realizing that they cannot control their time, these professionals feel uncomfortable about the loss of temporal autonomy and anxious about their inability to fulfill necessary family responsibilities such as childcare and eldercare (Mason 2020). In Kunming, Zhang (2020) also notices the ubiquitous anxiety about the loss of temporal autonomy. Some even struggle with mental distress and must resort to psychological intervention in order to live a better life in an anxious time.

In 2020, the debate on involution (neijuan, 内卷) in China is one example among many that reveal the ubiquity of social anxiety. The word involution was initially deployed by Clifford Geertz (1963) to describe social stagnancy and diminishing marginal utility in Indonesia's agricultural society. Without significant breakthroughs, the society spirals into a condition of internal friction, with increasing complexity instead of expansion. Involution has become a new buzzword in mass media in China to convey a general feeling of defeat when individuals realize that they cannot move up in the social ladder no matter how hard they work. Under the living conditions characterized by speed and efficiency, individuals have limited time to make their home, and become implicated in a strong sense of disorientation.

Second, the prioritization of housing over home further reinforces disorientation. In 1998, the central government in Beijing terminated welfare-based housing allocations and created a market-based system of housing provision, which kicked off the housing reform to foster the privatization and commodification of housing (Wang and Murie 2000). Since then, the house has occupied a central position in China's socialist market economy and evolved into a core package of personal property associated with child education, future financial security, social status, and economic return. The rush to house ownership is driven partly by capital investment and land-based finance in Chinese cities, and partly by the social need to gain recognition from others, including family members and friends (Lin 2014; Zhang 2012). Today, property ownership remains the primary standard by which to judge whether a person has the legitimate right to settle down in big cities.

The real estate market opens a Pandora's box by driving individuals to revolve their time around the purchase and sale of houses and forcing them to tie personal identity to the amount and status of property ownership. To many people, their life goal is to own a propertied house in a metropolitan city so that they can have a place in the city. To achieve this goal, they must work hard and do overtime in order to receive adequate revenues for a down payment and housing mortgage, a life-long commitment about which few can feel easy. For those who are fortunate enough to already own an apartment in Chinese cities, they desire for one more as an investment project to hedge against financial insecurity in the future. Ironically, this desire renders their daily life insecure as they worry about whether they can "keep up with the Joneses."

As property ownership becomes the foundation of social fame and economic wealth, individuals have no choice but to maximize the use of their time. In this race, money making replaces home making to serve as a primary force that organizes human life. In the apartment that they either rent or own, individuals are so busy with work and phone-based entertainment that they do not have adequate time or the appropriate mood to interact with their loved ones. Private property is supposed to foster freedom from work and public responsibility and create moments of leisure, comfort, and family care. In reality, it becomes an extension of the workplace or is overwhelmed by tedious household work. To make things worse, endless overtime pulls individuals out of their apartments so that their private life is disrupted. A house without personal feelings and social ties seems lifeless and meaningless. If there is a meaning, it is about social status at present and return on investment in the future. When the house is overwhelmed by rational calculation and hectic pace, it appears no different from a hotel room. The freedom that it confers is the freedom of financial transaction in the real estate market. Those residential apartments are carefully designed and professionally maintained, but how much time individuals invest in them for homely living is questionable. Their hectic schedule aims for making house, instead of making home.

The third and final reason accounting for loss of home feelings is the spatial alienation of everyday life in cities. To develop new high-rise apartment buildings for booming middle-class consumers, Chinese cities have become sites of massive redevelopment and relentless expansion, with bulldozers tearing down old buildings and giant tower cranes popping up to construct urban steel-and-glass forests. Old neighborhoods have disappeared, replaced by faceless steel towers or hypermodern high-rise buildings. Countless settlements such as urban villages and informal residential compounds continue to be bulldozed in the name of modernization, beautification, economic development, and urban imagineering. Accordingly, many Chinese cities displace longtime residents to the distant edges, and replace small, old houses with expensive high-rise apartments and office towers of spectacular design (He and Wu 2009; Zukin 2009). Ordinary residents have been dismayed by the way their cities have morphed from monolithic Maoist urban landscapes into new monolithic arenas of profit making and commercial transactions. Littered with exploitation and coercion, this transformation is stirred by conspicuous displays of wealth and power as private developers and city officials build for the rich, by the media who translate a dream of middle-class society into a frenetic party of consumerism, and by the gaudy tastes of the nouveaux riche and new urban middle classes who aspire to a modern, or even hypermodern, lifestyle whose primary purpose is to win the envy of others.

The forces of development and redevelopment for a booming real estate market have intensified social stratification, swept away the ties of community life and the traces of cultural tradition, and overtly scoffed at the idea that poor people have a right to live and work in cities. The right to the city is entangled in the continuum of organized violence. As shown in Shanghai, Beijing, and other Chinese cities, urban transformation, which has been planned by the authorities under the guise of economic competitiveness, involves "war-like levels of violence, destabilization, rupture, forced expulsion and place annihilation" (Graham 2004, p. 178). For instance, urban authorities in Beijing launched expulsions against low-end migrants in November 2017 by demolishing undocumented buildings and shattering unregistered factories (Zhan 2022). As a result, this aggressive campaign uprooted and displaced tens of thousands of migrants and left them abruptly houseless. One of the goals was to ease mounting pressure caused by urban sprawl and make room for high-end professionals with university degrees and good pay.

Rapid urban changes have disrupted routines and rhythms of everyday life and discouraged residents from feeling at home in their own city (Blunt et al. 2021). The discomfort of no longer feeling at home in once-familiar neighborhoods unsettles the spatialities and temporalities of home in the city. As Heller (1995) notes, familiarity is the most decisive constituent of the feeling of being at home.

Once people feel strange in their familiar place, they develop an uncanny feeling. In contrast to other scholars who separate the uncanny from the mundane, Freud (2003, p. 340) reiterates that "the uncanny is that class of the frightening which leads back to what is known of old and long familiar." In this sense, we cannot treat the uncanny as an experience foreign to individuals or reinforce familiar versus strange. The uncanny is frightening not because it is unfamiliar or new, but because what used to be familiar has somehow become strange. The familiar-to-strange continuum characterizes a mundane aspect of China's great transformation in which individuals must cope with a sense of strangeness in a familiar setting.

Realizing that urban redevelopment could dramatically change their familiar built environment, ordinary residents deploy various tactics to covertly confront developers and slow down the pace of change. For instance, residents in Beijing's old city launch legitimate complaints against developers so that they can defend their own interests. These complaints have emerged as "a disabling force" in the course of state-led spatial restructuring in Beijing, thus challenging the spectacular speed and telos of urban development (Nguyen 2017, p. 651). While Nguyen (2017) regards these site fights as a great challenge to the architecture of China's utopian futures, I emphasize that these elderly residents endeavor to defend a right to their home so that they can avoid the fate of spatial displacement and ensure a sense of home in their familiar neighborhoods. In Nanjing, the capital city of Jiangsu province, residents in old neighborhoods still live in traditional courtyard houses and enjoy close social relationships with neighbors (Tynen 2019). On the brink of demolition and displacement, these residents explicitly oppose being relocated to high-rise apartments, the symbol of modernization and happiness promoted by the local city government, but the epitome of loneliness and alienation in their viewpoint. They reiterate that "their homes in the old city, no matter how dilapidated, are still valuable to them" (Tynen 2019, p. 185). Despite the vernacular countermovement against spatial displacement in Beijing, Nanjing, and other Chinese cities, radical urban transformation has inevitably taken place and plunged ordinary residents into an unhomely condition.

In sum, modernization in post-Mao China entails an endless process of unmaking and making home. With some simplification, we can identify two explicit punches. First comes the transformation of house into an estate property, resulting in an endless race for wealth, and second is the outbreak of anxiety and insecurity regarding employment, education, childcare, and so on, to such an extent that individuals have no time and energy to build meaningful relationships with their surroundings. Zhang (2020, p. 137) describes the rise of middle-class anxiety in China's big cities:

> [M]any middle-class Chinese feel trapped and alienated by the world around them, a world that is supposed to be developing and improving materially—more roads, more high-rises, more cars, more housing, more food, more clothing. Yet, the sense

> of happiness, security, and contentment they have anticipated does not follow. Instead, they are troubled by confusion, emptiness, and uncertainty, despite living a materially comfortable life.

When the promises of modernization degenerate into the troubles of confusion and uncertainty, some fall into mental illness and some others feel lost in the race for wealth and fame. As Zhang (2018) points out, many urban Chinese are experiencing varying degrees of mental and psychological distress as a result of increasing market competition, dramatic social changes, and high work pressure. These social ills raise fundamental questions about the ultimate goal of modernization. From this perspective, the loss of home feelings represents a crisis of social transformation that entraps and alienates many in China's big cities.

Direct meticulous control prevalent in the Maoist period waned in the post-Mao era. A new mode of control, based on the rationalization of time and space, is exercised jointly by the state and by capital to shape the practices of everyday life. The race for wealth and fame, reified through property ownership, luxury goods, and social status, pushes individuals to comply with this new mode. Henceforth, the nature of time is transformed, from the quality of reflection and loafing in Chinese cultural tradition to the quantity of economic value and profit making. What Drucker (2001, p. 11) describes as the price of social advancement applies to China: "Upward mobility of the urban, affluent knowledge class comes at a high price—the psychological pressures and emotional traumas of the rat race." A loss of home feelings has emerged as a potent signifier of the disoriented and uprooted condition shared by individuals in China's big cities.

A loss of home feelings unfolds in the paradoxical time-space of modernity to implicate individuals into a permanent search for homely life, a search spelling out "an almost paranoid and painful desire" (Kraftl 2007, p. 136). It is paranoid because individuals long for the not-yet, or utopian imaginaries, and painful because this endless journey is imbued with hope and disappointment. The booming market economy in post-Mao China enables Chinese people, particularly those in big cities, to accumulate pecuniary ability to pursue their favorite lifestyle. This enabling factor is best reflected in China's thriving tourism industry (Su and Teo 2009), millionaire emigration (Liu-Farrer 2016), and psychological training for self-development in cities (Zhang 2018). People begin to realize that getting rich does not necessarily guarantee a deep sense of fulfillment and happiness. Some of them develop a taste for alternative temporalities. For instance, a new counseling movement has emerged in Chinese cities to promote psychological training and interventions for the goal of achieving "a stronger sense of individual autonomy and well-being" (Zhang 2018, p. 54). Meanwhile, some people travel elsewhere and spend much time and money in preserving health (*jiankang*, 健康) and wellness (*yangsheng*, 养生).

A search for home via alternative time-spaces denotes lifestyles different from speed and fury. By trading money for meaning, this new deal of modernity may bring about the possibility for healing and loafing that are desperately wanted in a country that has undertaken a great transformation with such a high speed. The new taste for alternative time-spaces elsewhere can be understood as "both responding to and in some sense enabling some form of 'retreat' from the 'ravage' of time-space compression" (May and Thrift 2001, p. 20). How to understand the entanglement between the loss of home feelings and the search for home requires a critical analytic framework. Next, I will elaborate this framework on the basis of unhomely life.

Unhomely Life: An Analytic Framework

Home and homelessness in the contemporary order of modernity always represent two logical possibilities. The first is that a homely life is experienced to make individuals feel at home, not necessarily in relation to a housing property, but more in relation to subjective meanings such as belonging, comfort, and security. Such feelings constitute a resistance to the temporal and spatial order of modernity and preserve ways for individuals to counter alienation, discipline, and even commodification. The second possibility entails a loss of home, or a sense of homelessness, another order of modernity. This sense of homelessness is not associated with property ownership or national citizenship, although scholars and the media constantly misuse *homeless* to describe those who live on the street and have no stable shelter or property in which to dwell. I call the latter *houseless*.

Different from houselessness, or a lack of decent shelter for dwelling, homelessness here refers to a general condition in which individuals lack a feeling of being at home, either representationally or experientially. Tucker (1994, p. 184) succinctly summarizes homelessness as a state of "utter social loneliness." It is an experience of rootlessness or discomfort in a world relentlessly moving towards alienation and stratification. Modernity ascribes a certain sense of homelessness to modern subjects. To Heidegger (1962), homelessness is coming to "be the destiny of the world," while Berger et al. (1974) emphasize the spreading condition of homelessness and the rise of "homeless mind," all pointing to a crisis of home. If home is in crisis, it is not because the custodians of the traditional family – women – have left its bounds to enter into the paid workforce, as Duyvendak (2011) proclaims, but because individuals, regardless of their gender, ethnicity, and class, lose the cultural or natural roots that they inherited from previous life journeys and miss the beacon of hope and inspiration for future journeys.

To withstand homelessness, some resort to nostalgia, some pursue idyllic lifestyles in the countryside, and others choose to live with it and develop a feeling of being at home everywhere. The question of homemaking is a question concerning how individuals mitigate "the fear of homelessness generated by the hustle and bustle of modernity" (Morley 2000, p. 43). The question is not merely about being – how individuals stand in relation to their self, but also about living – how they negotiate for an acceptable way of life. Rather than emphasize the distinction between home and homelessness, or attempt to analyze the cultural politics of either, we need to overcome the formulation of dichotomy and recognize that home and homelessness are inextricably interwoven to foster a status of being unhomely as a common expression of existence in a rapidly changing world. Home gives shape to homelessness, and vice versa, so much so that unhomely life remains a constant reality in the seesaw of modern urban life.

Seesawing between home and homelessness is unhomely life, an in-between condition common in the urbanized modern world. It entails felt-sense experiences ranging from dwelling in displacement (Clifford 1997) to homelessness at home (Bennett 2011), and a sense of community among homeless persons (Christensen 2013). The popular term "no place to call home" exemplifies an unhomely condition bereft of emotional attachment and ontological security (Giddens 1991). For instance, Blunt et al. (2021) find that residents on and near Kingsland Road, in east London, express the discomfort of no longer feeling at home due to dramatic changes in their neighborhood, and endeavor to cope with their changing surroundings. Once their familiar neighborhood becomes strange and inhospitable, these residents feel both at home and not at home, an unhomely feeling that encompasses "intertwined and sometimes ambivalent experiences, memories, and narratives of home, belonging, and displacement" (Blunt et al. 2021, p. 156). Thus, Brickell (2012, p. 227) calls for (re)visioning "a more complex and fluid understanding of the home which emerges from deliberate or unintentional disruptions to home places and sentiments at different scales and times."

Brickell's argument reminds us of Bhabha's (1994) cue regarding the process of displacement and disjunction that afflicts people, challenging normative narratives of identity and dwelling. Studies of unhomely life frequently refer to Bhabha (1992, p. 141) who uses the term "the location of culture" to describe "a tenebrous sense of survival" when people live on the borderlines of the present. Border lives, as Bhabha expresses to describe an in-between existence, entail the moment of transit where space and time cross to generate complex figures of inclusion and exclusion. As Bhabha (1994, p. 2) explains further, these in-between spaces create the terrain for explaining strategies of selfhood – either individualistic or communal – that produce "new signs of identity, and innovative sites of collaboration, and contestation, in the act of defining the idea of society itself." Bhabha couches the in-between location of culture as a protest against the structuralist approach to fixity in time and space, and applies this perspective to explore cultural transformation in postcolonial settings. Starting with classic literature and following the insights from Fanon and Said, Bhabha emphasizes that cultural transformation entails ongoing struggles between external influence and local contexts that yield neither once-and-for-all victories by Western colonialists nor exclusive nationalist sentiments. Once the organic process of transformation in indigenous lands was disrupted or displaced by

European influence, cultural modalities were juxtaposed or swung between two poles: the indigenous and the European. Thus Bhabha (1994) argues that the intrusion of the West into the indigenous home forged the eeriness of a familiar place, or a form of cultural hybridity, fostered through interactions and conflicts, to such an extent that the colony can neither go back to its origin nor become fully Westernized.

Bhabha further explains the existence of cultural hybridity in both European and postcolonial societies. To Bhabha (1994, p. 64), the distinctive characteristic of postcoloniality in the West and the Global South is "the form of otherness, the tethered shadow of deferral and displacement," which created uncanny dimensions of experience among the colonializing Self and the colonized Other. Each side relies on the other to construct their identity, either through imitation or orientalist reproduction, to mimic each other. As Bhabha (1994, p. 122, original emphasis) contends, colonial mimicry is "the desire for a reformed, recognizable other, *as a subject of a difference that is almost the same, but not quite*." His ideas on mimicry between the self and the other are forerunners of later work in postcolonial studies, showing the interrelation between these two sides. Seen from the perspective of the other side, what they are familiar with in their own setting turns into strangeness.

Indeed, Fanon (1970) uses the term mimicry to describe how the colonized elite attempted to emulate the colonizer's language, dress, and other cultural codes in order to receive recognition. Although these colonized subjects realized that colonialism robbed them of the humanity that they had once enjoyed during the precolonial period, they did not yearn for an essentialized precolonial society but for the internalization of colonial culture in an attempt to "regain a humanized 'home'" (Tinsley 2020, p. 2347). Ironically, the representation and imagination of a precolonial pastoral society by European cultural elites reified European nostalgia for premodernity and projected the notion of home onto the empire. While Europeans romanticized colonized societies as rural and exotic to fulfill their imagination of ideal home elsewhere, some colonized elites emulated and internalized European colonial culture so that they could reconstruct a sense of ideal home, again, elsewhere (Bhabha 1984). Both sides got stuck in an unfulfilled journey of homemaking with a hope to cultivate what they wanted, either through a loss of home due to modern industrialization or a dream for a new home due to self-negation. They participated in the deal of modernity: trading meaning for power. Because modernization is powerful enough to subordinate humanity and meaning to the logic of capital accumulation, the imagination of a premodern exotic society cannot secure a sense of home for Europeans. Similarly, mimicry as a route to new home by colonized elites is doomed to fail, because "colonialism is an inherently dehumanizing enterprise" to plague the colonized into an exile in their own land (Tinsley 2020, p. 2347).

As a result, postcolonial societies cannot return to their precolonial traditional formats, nor can they reach the benchmark of Western civilizations. They are stuck in an in-between space, which may be "contingent to [Western] modernity, discontinuous or in contention with it, resistant to its oppressive, assimilationist technologies" (Bhabha 1994, p. 5). Nevertheless, postcolonial societies are not passive recipients of Western modernity, but deploy the strategy of cultural hybridity to translate and reinscribe the social imaginary of modernity, giving rise to what Bhabha (1994, p. 13) calls "the rite of extra-territorial and cross-cultural initiation." Having said this, Bhabha cautions against any romantic notion of cultural diversity, arguing that the postcolonial passage through Western modernity generates a perpetual condition of cultural dislocation and intends to erase "that compliant past tethered to the myth of progress" (Bhabha 1994, p. 363).

The home, which is conceptualized by Bhabha (1994, p. 12) as a life-world "constituted by relatively stable associations, relatively known and shared histories," stands at odds with many people's life experiences. If an ideal home represents the first space of home, which entails a transformation from womb and parental houses to a new home with meanings such as comfort and familiarity, then we are torn apart by and fearful of homelessness, both spiritual and bodily. Homelessness points to the second space of home, another pole in our life journey, where individuals are implicated in rootlessness. Homelessness involves feelings of being lost, or what Bhabha (1994, p. 13) describes as "the estranging sense of the relocation of the home and the world." Between these two spaces lives the third space of home, conceptualized as a contradictory and ambivalent space of enunciation in which "the meaning and symbols of culture have no primordial unity or fixity" (Bhabha 1994, p. 37). Bhabha emphasizes that discursive enunciation can appropriate, translate, rehistoricize, and even reinterpret the same signs for certain purposes. For instance, the meanings of colonial inheritance can be transformed into "the liberatory signs of a free people of the future" (Bhabha 1994, p. 38). The in-between space, the site of engagement and negotiation, carries the burden of fostering unhomely life. According to Bhabha, the unhomely is a paradigm of postcolonial experience through which individuals relate the traumatic ambivalence of a personal, psychic history to the wider disjunctions of political existence and displacement. The unhomely condition fosters an estranged sense of encounter between the private and the public, familiarity and strangeness, and intimacy and alienation.

I believe that Bhabha's analytic framework has a broad application beyond postcolonial settings. Indeed, Bhabha (1994, p. 13) emphasizes being unhomely as an uncanny condition not confined to postcolonial societies: "Although the 'unhomely' is a paradigmatic colonial and post-colonial condition, it has a resonance that can be heard distinctly, if erratically, in fictions that negotiate the powers of cultural difference in a range of transhistorical sites." His perspective upon the unhomely as an in-between existence can provide a fruitful overarching framework for understanding the making and unmaking of home in a mobile world. Unhomely life entails the condition of transit where space and time cross to generate complex figures of inclusion and exclusion. Following Fanon's work on mimicry and Freud's work on the uncanny, Bhabha (1992, p. 144) argues that the unhomely epitomizes a temporal moment of transit in which the framing and naming of social reality is negotiated: "not what lies inside or outside reality, but where to draw the 'meaningful' line between them." And yet, unhomely life enfolds in "a tenebrous sense of survival, living on the borderlines of the 'present'" (Bhabha 1994, p. 1). Precisely out of such experiences of negotiation between personal life and structural forces (particularly in a period of radical transformation) do schemes of imagination and practice become inscribed and reconfigured. With the notion of the unhomely, Bhabha opens a path to a nonessentialist account of sociocultural transformation, one that is not fixed in one existence or uprooted to become a new existence. Rather, it articulates in-betweenness as a dynamic form of being-in-the-world, neither here nor there, but both here and there. Unhomely life itself is not a fixed modality, but unfolds in different contexts and moments to reflect and reproduce an in-between condition of living between here and there, across different temporal ranges (the past, the present, and the future), and with a mixture of home and homelessness.

Bhabha's analysis offers crucial insights into a variety of other cultural experiences beyond the postcolonial one that he addresses. Modernity and its temporal and spatial logic have made life neither homely nor homeless. There is no such state of existence that is completely "at home." Nor do individuals give up the pursuit of home, simply because a sense of home is so dear to human beings that efforts are made to cultivate such a sense no matter how complicated and strange the world becomes. For most individuals, somewhere in between becomes the position of living with modernity, and how to handle this in-between existence reflects the negotiation between themselves and the external world. For instance, bell hooks explains her loneliness in New York: "one lives in close proximity to so many people engaging in a kind of pseudo intimacy but rarely genuine making community" (hooks 2009, p. 24). Bennett's (2011) work on the experiences of young working-class women living in privately rented or public housing in east Durham in the United Kingdom is another indication of unhomely life, though the author chooses the term "hidden homelessness." As Bennett describes, homelessness at home comes from "the spectral presence of 'others' in the home" as these young women have constantly encountered social workers, police officers, or property owners. The ignominy of "hidden homelessness" experienced by these women as dispossessed and haunted has been the subject of other research, including the work of Miller (2001), which provides an insightful account of unhomely life in publicly owned apartments in London. Such homelessness amounts to living with "fears and frustrations" at home (Miller 2001, p. 119). This sense of unhomeliness similarly happens to mobile migrants in Hong Kong who, according to Jankowski (2018), get stuck in between here and there, or between two extreme poles – the vagabond and the tourist. Though Jankowski uses the term "middling mobility" to describe this in-between situation, these migrants engage in an unhomely life because Hong Kong becomes a site of temporary living, no matter how long they stay in the city.

To understand this in-between condition of unhomely life, I propose a journey-oriented perspective. One journey involves leaving a previous home to explore "a world of possibilities" (Lefebvre 2002, p. 6), dwelling with belonging and alienation in the current home, and imagining a new home in the future. Wardhaugh (1999, p. 94) reminds us of the importance of life journeys by arguing that "a 'home' with no 'journey' becomes a prison, in that the home may serve as a source of order and identity, but with no opposite, no 'outside,' the individual becomes constrained and confined." The home, as an embodiment of ideal existence, may never be fully identified or reached, but motivates individuals to launch journeys in search of a better life for themselves or the next generation. During these journeys, some choose to reconcile with the reality and accept whatever they have so that they can cultivate a settled sense of home; others endure the uprooting condition here and now, and still others refuse to give up the ideal but pursue a new home elsewhere, temporarily or permanently. These variegated choices highlight that homemaking reflects and reshuffles the dynamic relation between the self and the world. In the process of making and unmaking home, the journey is filled with continuity and disruption. As Tucker (1994, pp. 184, original emphasis) points out, "most people spend their lives in search of home, at the gap between the *natural home* and the particular *ideal home* where they would be fully fulfilled." Homemaking at one moment is supposed to address this gap, but might generate new ideals of home for the next journey. Thus, a journey-oriented perspective understands daily life as neither a static form of being nor an endless condition of becoming, but entails observable patterns of continuity and disruption when people move from one place to another, from one time to another.

This journey-based approach, in its temporal (past, present, and future) and spatial (here and there) dimensions, can help us to conceptualize homemaking as a continuous process of negotiation, compromise, and resistance (Mallett 2004). This approach is salient for analyzing mobile subjects in Lijiang, as people on the move converge in the town and desire for better living, no matter how they define it, culminates in the process of making and unmaking homes between Lijiang and elsewhere. The town acts as a bus stop (Heller 1999), or a contact zone (Pratt 1992), to accommodate their life journeys for a while and stage their struggle against and compliance with speedy modernization. What may seem the one-way pathway of hollowing out modern life is still open to other possibilities, as individuals gauge the cultural and economic tenor to negotiate for a better way of life pegged more to meanings than profit. In the following chapters, I will turn to Lijiang and explore dynamic pictures of making and unmaking home by four groups of people: lifestyle migrants, retreat tourists, native town residents, and local service workers.

Notes

1 Regarding the theory of everyday life, Lefebvre (1991a) emphasizes the relationship between the individual and the social as essential and critical. I expand it to the relationship between the self and the world in order to highlight the importance of two essential features of nature – inner and external – to homemaking.

2 《论语·为政》(七十而从心所欲，不逾矩), translation comes from *China Daily*, source from https://language.chinadaily.com.cn/a/201809/28/WS5b20a940a31001b825720b5f.html

3 Volition is a key feature of autonomy in Kantian moral philosophy. While Kant elaborates the moral aspect to highlight the importance of autonomy of the will, I focus more on personal autonomy and how individuals manage time and space in their everyday life. For more on the Kantian notion of autonomy, see Darwall (2006) and Shell (2009).

4 Saunders and Williams (1988) note the difference between house, home, and household, pointing out that home is a physical setting through which to constitute and reproduce basic forms of social relations. In other words, home represents a socio-spatial system to fuse the physical unit or house and the domestic unit of household. While I recognize the value of this differentiation, I attempt to detach home from house and household in order to explore the cultural construction of home. The term home can be invoked in various contexts, such as homeland, home culture, nursing home, and homely life, in which the notions of household and family do not apply.

5 Mallett (2004) summarizes some critics of the conflation between home and family, pointing out that this conflation is ideologically laden and draws on the white, middle-class, heterosexual nuclear family.

6 The full text of the speech can be found at: https://www.margaretthatcher.org/document/105815

7 The full text of the speech can be found at: https://georgewbush-whitehouse.archives.gov/news/releases/2003/12/20031216-9.html

8 https://www.tandfonline.com/action/journalInformation?show=aimsScope&journalCode=rfhc20

9 A house has no cultural meaning in itself. Or as Le Corbusier famously argued, a house is a machine to live in. Technically, it has only commercial and functional value, but it can become a home only when meanings and feelings become attached to it (Easthope 2004).

10 This unity is akin to Heidegger's understanding of dwelling, which entails "the capacity to achieve a spiritual unity between humans and things" (Harvey, 1996, pp. 300–301). Indeed, Heidegger (1962) invented the term Dasein, or being-in-the-world, to emphasize that people exist only through their relationships to the world. Similarly, Tuan (1990) coined the term 'topophilia' to conceptualize the effective bond between people and place. Having said this, the Chinese idea of nature–human unity entails some key cultural connotations to understanding the dynamic relationships between the self and the world.

11 Source from: 庄子，外篇，达生(Outer chapters, The Full Understanding of Life). Translated by James Legge, source from: https://ctext.org/zhuangzi/full-understanding-of-life/ens

12 A similar conception in the West is that going home means searching a safe haven in a heartless or tumbled world (see Duyvendak 2011; Lasch, 1977). This conception can easily deduce nationalist sentiments to exclude those who might jeopardize the haven.

13 Original text comes from:白居易《初出城留别》:

朝从紫禁归，暮出青门去。勿言城东陌，便是江南路。
扬鞭簇车马，挥手辞亲故。我生本无乡，心安是归处。

In this poem, Bai Juyi describes a dramatic change in his career, returning from the imperial palace in the morning and being exiled from the capital Chang'an in the afternoon. Waving farewell to friends and relatives, Bai must accept the penalty and travel to a prearranged destination. Reflecting on this political and spatial change, Bai Juyi notes that where there is ease is his home destination. In a different poem, Bai Juyi writes an identical lyrics by saying that regardless of where I am, home is where I feel ease (无论海角与天涯，大抵心安即是家). The full text is as follows

种桃杏
无论海角与天涯，大抵心安即是家。
路远谁能念乡曲，年深兼欲忘京华。
忠州且作三年计，种杏栽桃拟待花。

Similarly, Su Shi (苏轼), a literary giant in the Song Dynasty (960–1279), records the same expression when he asked a woman, who accompanied an exiled cadre, where is her home: 此心安处是吾乡(Where there is ease, there is home). Scholars in exile in ancient China used the phrase "Where there is ease, there is home" to console themselves. Some prestigious scholars in contemporary China also use "Where there is ease, there is home" as their book title to reflect on their versatile life journeys, including Yu Guangzhong (余光中) and Ji Xianlin (季羡林). The conception of home as where the heart is at ease remains a cultural genre in China.

14 Translated by Hinton (2005)

15 In Chinese, 久在樊笼里，复得返自然, translated by Hinton (2005).

16 A reclusive life experienced by Tao and others is different from life as an ascetic hermit. Recluse means living a civilized life with necessary material processions such as books, decent dwelling, and food and drinks. It does not reject family life, friendship, and daily socialization with farmers. In contrast, a hermit lives a very simple and frugal life, generally alone, so that he or she does not socialize with others beyond what is necessary.

17 The title is Q&A in mountains (山中答问),Translated by Seaton and Maloney (1994, p. 48)

18 Lin (1998) regards the knowledge of truth as the eternal focus of Western philosophy so as to pay inadequate attention to the knowledge of living itself. Seeing that rational

logic overrules real living in the West, Lin (1998, p. 160) jokes that the German philosophers are the most frivolous of all: "They court truth like ardent lovers, but seldom propose to marry her."

19 In Chinese 道可道，非常道 (translated by James Legge, source from https://ctext.org/dao-de-jing/ens).

20 "百日之蜡，一日之泽，非尔所知也。张而不弛，文武弗能也；弛而不张，文武弗为也。一张一弛，文武之道也" (礼记.杂记下); translated by James Legge, and source from https://ctext.org/liji/za-ji-ii/ens. King Wen and King Wu, two famous rulers in the Zhou Dynasty (510–314 BCE), were regarded as role models for kings and emperors throughout Chinese history.

21 Original text: "人生而静，天之性也；感于物而动，性之欲也" 《礼记-乐记》 (Confucius, 1967)

22 Scholars raise various interpretations of home as haven or heaven in the European tradition. Both metaphors allude to Christianity to portray an ideal afterlife.

23 https://dengxiaopingworks.wordpress.com/2013/03/18/there-is-no-fundamental-contradiction-between-socialism-and-a-market-economy/

24 http://www.chinadaily.com.cn/a/201904/16/WS5cb52e18a3104842260b6625.html

25 Source from http://www.china.org.cn/living_in_china/abc/2009-07/15/content_18140508.htm

Chapter Three
Lifestyle Migration and the (Un)making of an Ideal Home

The importance of lifestyle cannot be underestimated in a consumer society. People are concerned with the stylization of life, or with "customizing a lifestyle and a stylistic self-consciousness" for the purpose of individualistic ethos (Featherstone 1987, p. 59). Featherstone (1987) further contends that the subjects of consumer culture make lifestyle a life project by displaying their individuality and sense of style through an assemblage of goods and experiences. Meanwhile, lifestyle is akin to what Bourdieu (1984) calls distinction, namely, a way of expressing and identifying social groupings through taste. Today, the term points to a recognizable and even distinctive way of life, which is rooted in various social and cultural elements such as music, food, dwelling, and clothing. Among the various ways of pursuing a lifestyle, migration is singled out (Åkerlund 2017; Benson and Osbaldiston 2014; O'Reilly and Benson 2009). To Benson and O'Reilly (2009, p. 609), lifestyle migration refers to "the spatial mobility of relatively affluent individuals of all ages, moving either part-time or full-time to places that are meaningful because, for various reasons, they offer the potential of a better quality of life." Lifestyle migrants often frame and even idealize their destinations as offering various possibilities to engage in individualistic ethos and self-expression, an idealization themed by "escape" or "utopia" (Benson and O'Reilly 2016; Benson and Osbaldiston 2014; Thomas 2008).

The term "ideal" is in part metaphorical, as it points to the representation and construction of particular destinations as an ideal or, more often than not, idyllic, home (Benson and Osbaldiston 2014). Indeed, no matter how scholars interpret

Unhomely Life: Modernity, Mobilities, and the Making of Home in China, First Edition. Xiaobo Su.

its nature, the home in its metaphysical form always denotes some positive feelings. As Blunt and Dowling (2006, p. 100) argue:

> A central feature of imaginaries of home is their idealization: certain dwelling structures and social relations are imagined to be "better," more socially appropriate and an ideal to be aspired to. It is these dwelling structures and social relations that become "homely homes."

This emphasis on "better" and "homely" involves a dynamic process of idealization, a process in which individuals can enjoy spatial and temporal autonomy and cherish "the cultivation and nurture of the individual self" (Conradson and Latham 2005, p. 292). In the case of lifestyle migration, being ideal can be associated with "authenticity, implying simplicity, purity and originality" (O'Reilly and Benson 2009, pp. 4–5), the desire for solitude and isolation (Carson and Carson 2018), nostalgia and romanticism (Buller and Hoggart 1994), or the "rural childhood idyll" of safety and closeness to nature (Baylina and Gunnerud Berg 2010, p. 287). Thanks to their "ideal" (either constructed or imagined) nature, destinations can be endowed with various meanings, such as a sense of community, self-fulfillment, and place attachment, and thus, become a new home. The idealization of destinations provides interesting insights into how lifestyle migrants attempt to make a new home.

Different from those in the position of disadvantaged mobility, lifestyle migrants engage in a privileged mode of mobility. In general, they seek "satisfaction from aspects other than work: good weather and food, better housing, new spiritual or cultural experiences, or a more relaxed, less materialistic culture" (Leonard 2010, p. 2). The pursuit of a better quality of life is neither abstract nor metaphysical. It crystalizes into the process of making an ideal home articulated through material elements and daily practices. Once the pursuit is jeopardized in destinations, lifestyle migrants may choose to unmake home, that is, leave for a new destination for another ideal home or return to their original place. This process of making and unmaking illustrates that lifestyle migrants face a unique set of struggles and tensions from here to there. Therefore, home is better understood through translocality and multiple temporalities, and conceptualized as "a metaphorical and potentially multi-sited space of personal attachment and identification" (Brickell and Datta 2011, p. 24). This condition of translocality thrusts mobile individuals into an in-between condition of both here and there, and neither here nor there.

This chapter highlights lifestyle migration as a new form of mobilities in contemporary China. Notwithstanding relatively smaller flows of people, its effects can be exemplary and far-reaching for two reasons. First, the insistence on favorite lifestyles runs counter to the mainstream trend of migration in China, indicating an important aspect of social transformation emerging from China's speedy modernization. Second, some people can escape from the manacles of time discipline to enjoy favorite lifestyles, highlighting the working of human agency to mitigate social acceleration in China. Lijiang is one such popular destination. In the past two decades, tens of thousands of lifestyle migrants have left China's big cities for Lijiang and integrated business and leisure into rented properties for themselves and tourists as well. To them, lifestyle is about a balance between meaning and money. Meanwhile, these migrants keep a distance from Lijiang's local natives so that they can avoid daily trivial and external constraints. By doing so, they embrace and enjoy an idealized way of life under the shadow of social alienation and government regulation. Once the external condition deters their efforts to maintain the imagination and experience of ideal home, they choose to unmake their home by leaving Lijiang.

Representing Lijiang as an Ideal Home

The making of an ideal home by lifestyle migrants in Lijiang can be characterized by either homesickness or evasion. It is homesick because migrants idealize destinations and the ways of new life that they can enjoy with more comfort and less stress than usual. Such ways of life are usually associated with a good balance between work and relaxation, a sense of community, or closeness to nature. The making of an ideal home also is evasive because migrants want to escape the increasing temporal and spatial constraints imposed on their daily life by capitalist modernity. The purpose is to cultivate "a recreation, restoration or rediscovery of oneself, of personal potential or of one's 'true' desires" (O'Reilly and Benson 2009, p. 3). Here I concur with Blunt and Dowling (2006, p. 119) who argue that idealized views of home are important to "a geographical understanding of home." While Blunt and Dowling (2006) mainly focus on the social construction of the middle-class suburban house as an ideal home in contrast to city dwelling, I place the ideal home in the context of lifestyle migration to illustrate the intertwining of practices of dwelling (belonging and routine) and mobility (escape and adventure).

Lijiang Old Town gained the spotlight for lifestyle migration once pioneering middle-class professionals chose it in the early 2000s. When I visited Lijiang to conduct my fieldwork in 2002, I noticed that migrants operated nearly half of the tourism-related shops along the main streets, and some of them came for Lijiang's exceptional living conditions, which include unique ethnic culture, ancient streetscapes, and pleasant natural amenities. The inflow of lifestyle migrants has continued since then. As one well-known guesthouse operator in Lijiang proclaims in her blog:

> [It is about] busy metropolitans vs slow-pace towns! Traveling is easier than settlement. With regard to the meaning of old streets in this small town, Lijiang to China is on a par with Provence to France, and Hallstatt to Austria. It is not merely

> a tourist destination where people can wander around or a cultural heritage shared by the whole world, but also a site for alternative lifestyle. (https://m.weibo.cn/detail/4249108073589461, Author's translation)

This statement is a typical example of how Lijiang is idealized as an attractive site where people choose a different, and often more satisfying, way of life. It highlights that the ideal home is embodied as a geographical imagination of alternative lifestyle that these migrants desire. Similar narratives and mythologies of staying in Lijiang become intertwined with ideas of the alternative lifestyle.

The pursuit of an alternative lifestyle, quite different from the hustle and bustle of urban living in China's big cities, underpins the respondents' imagination of an ideal home. The story of one couple is exemplary. Xue (female, a guesthouse operator, from Shanghai) came to Lijiang in 2014 after one of her friends took over a guesthouse and asked whether she would like to take a break from hectic urban life in Shanghai. As she described during the interview, "Lijiang is my dreamland. Many TV dramas focus on the town and plant the seed of a dream in my mind." With the invitation, she visited Lijiang for a break. After a while, she decided to resign from her job in Shanghai and chose to stay in the guesthouse and help the owner to run it so that she could enjoy the Lijiang lifestyle for as long as possible. Later, Xue and Wang (male, a guesthouse operator, from Beijing), fell in love, got married, and settled into life in Lijiang. Talking about their earlier lives in big cities such as Shanghai, Xue lamented that "If both of us (Xue and Wang) lived in cities, we would always feel anxious and nervous. Why? It is because of high pressure. Of course, we had pressure in Lijiang in the past two years, due to the property rent or room occupancy. Gradually we collaborated with each other to establish our business."

Wang's story is similar to Xue's. Before coming to Lijiang, Wang was a real estate manager in Beijing with a very busy life. After a series of personal changes, including a breakup with his ex-girlfriend, he felt exhausted. So he resigned and traveled in Yunnan and Tibet, hoping to reflect on his personal life and open his mindset about what he wanted to do next. He spent half a year traveling to keep a distance from hectic schedules and to calm down. The transition from a hustling lifestyle in Beijing to a relatively comfortable slow-paced life was an uncanny experience for Wang, as he had to retrieve the meanings of time and space from what he encountered during his journey. After a careful market survey, he decided to take over a guesthouse in Lijiang in 2014 and start a new life journey with a permanent farewell to Beijing. Here is a conversation about the sea change between Beijing and Lijiang:

Author: What do you think of your life?

Wang: Our life is related to the environment and our attitude. Attitude is particularly important. People in big cities are bereft of inner satisfaction, no matter how rich they are. Some of them even suffer from mental illness, although they don't want to face these problems.

Author: How is your life after migrating to Lijiang?

Wang: The biggest change is the quality of life. In the past I believed that you could enjoy a certain style of life only after you owned a good materialistic foundation. Nevertheless, we don't know how much is enough. When I traveled in Yunnan and Tibet, I found I was wrong. Our economic gain has already met the bar, but our consciousness lags behind. Lijiang is a place that is inclusive and open. It contains various new thoughts and possibilities, making it different from big cities.

Wang's reflection on the relationship between meaning and money, or the deal of modernity, is commonly shared by many other lifestyle migrants in Lijiang. Rather than following the style of reclusion in ancient China, which prioritized meaning over money, Wang emphasizes the need to strike a balance between these two elements.

These two stories illustrate that lifestyle migrants from China's big cities are actively involved in making a new home through a different experience of time and space, with strong attempts to reject time discipline. Indeed, distinctive attitudes often exist regarding temporalities here and there, since they compare Lijiang's tempo to their experience before. Here Wang exemplifies the enjoyment of loafing that most lifestyle migrants in the old town expressed to me on multiple occasions. This temporal difference further justifies their choice of Lijiang for settlement and pushes them to maintain such a homely milieu for both themselves and tourists. After all, the selling of an ideal home experience to tourists proves to be a key business strategy (see Chapter 4). The old town has been transformed into an oasis of loafing with new cultural meanings, developed by lifestyle migrants, for the purpose of personal enjoyment and tourist consumption. Wang also alludes to the great affliction brought by speedy life and time discipline in big cities such as Beijing and expresses dissonance with the dominant discourses of modernization. Although he did not specify who should account for the affliction, he is content with his new life in Lijiang. The making of ideal home in Lijiang is an example of the spatial production of lifestyle as an alternative to speedy modernity.

If retrieving the meanings of life could be too abstract to others, a recovery from mental and bodily fatigue remains a key incentive for many lifestyle migrants for settling down in Lijiang. Before migrating to Lijiang in 2014, Xiao (male, from Xi'an) was a music producer and operated a small studio in Xi'an. He was absorbed in his work to such an extent that his physical and mental health started to deteriorate. He said that he usually worked fourteen hours per day, seven days a week, and inevitably his health suffered to the point that his body could not bear high-pressure work anymore. Henceforth, he decided to stay in Lijiang for a different lifestyle. At the time of our interview, Xiao worked as a music director in a night bar. To him, Lijiang is an ideal place to dwell with comfort and to relax with fun:

> I have always loved Lijiang for many years. I came to Lijiang not for music, but for a different lifestyle. I had to escape my previous life. Although I earn less in Lijiang than before, I do not prioritize work over life. In Lijiang, I enjoy myself much better than before. The most important things in Lijiang include drinking tea and enjoying sunlight, normally in the yard I rent.

For Xiao, the alternative lifestyle in Lijiang provides him with a sense of an ideal home in contrast to his otherwise frenetic pace of life in Xi'an. Through this comparison Xiao expresses a desire for and fascination with Lijiang by choosing to settle down and start a new life, which demonstrates an enduring longing for an ideal home where he can enjoy a better quality of life. This enjoyment reflects two aspects of the ideal home in Lijiang. First, the ideal home is tied to a better quality of life as individuals have more autonomy over their time and space. Second, migration to Lijiang represents an escape from big cities and a departure from the past for the sake of an alternative lifestyle elsewhere.

This departure from the past is similarly narrated by Mei (male, from Shanghai). In Shanghai, Mei had worked in different firms for nine years. One common thread that ran through his journey in Shanghai was constant busyness. As he complained, he did not want such a rapid and busy pace of life with such high pressure. Haunted by moments of burnout, he started to search for an ideal place and eventually chose Lijiang as a place to settle down in 2015. Mei's main business is to operate a tourist guesthouse, which, he says, is easy and simple. While money is not a major concern, he cherishes the opportunity to enjoy more leisurely time and have fun with tourists and other migrants. To him, the guesthouse is his home where he can live a comfortable and relaxed life. In another case, Lin (female, from Dalian) came to Lijiang in 2010. Her stay in Dalian disappointed her, partly because she rushed to do things not in her interest, and partly because she did not become successful as expected. She then resigned and traveled to Lijiang for the Chinese New Year in 2010. Soon she decided to stay and try things she loved. To her, Lijiang Old Town is a place to calm down and pursue an ideal lifestyle characterized by simplicity and comfort. As she puts it, "the town has a very inclusive cultural atmosphere and I feel quite comfortable. I rarely travel to other places because I am not a tourist. I live in this town."

Stories of this kind abound, repeated many times by other lifestyle migrants I interviewed. An analysis of their narratives about the unmaking of their home in big Chinese cities can help to illuminate what is at stake for lifestyle migration in Lijiang. They pursue a lifestyle radically different from speedy modernization in China's big cities (Su 2014; Wu and Yang 2022). In particular, this tempo in big cities that my respondents describe is contrasted with the slow, cyclic, and habitual rhythm of rural life that Fei Xiaotong (Fei 1992) describes in Xiangtu China. This contrast is further reinforced by other features of social and economic psychology in big cities, once individuals get used to the calculating attitude derived from the money economy and the ubiquity of anonymous encounters.

Thus, migration to Lijiang seems to generate a new and rewarding sense of comfort and loafing. The comments of Xiao and Mei already reveal some of the answers. Similarly, Lily (female, from Harbin, migrated in 2015) runs a café and talks about the difference in her life: "The difference is obvious. In the past, I had to get up at 5:00 a.m. to take care of two kids. Then I ran to my office and then to my kids' schools. What a hectic pace of life! After I came to Lijiang, I now wake up so naturally that I rarely get up earlier than 10:00 a.m." This sort of change also happened to Deheng (male, from Shanghai), who quit his job with a Fortune 500 corporation and migrated to Lijiang in 2011: "In the past, I received 200 to 300 calls per day. Can you imagine how busy I was? Now my phone rings no more than 30 times in one month. This drop in the number of phone calls reflects the change in my life." For both Lily and Deheng, the pursuit of a better quality of life is associated with a slow, low-pressure pace and lighter workload so that they can calm down for relaxation.

Lifestyle migrants in China relocate to Lijiang and other small cities to find the ideal home that speaks of alternative lifestyles. Xu and Wu (2016) attribute this relocation to the emergence of a new middle class, the arrival of a consumerist society, and traditional Confucian culture. These enabling factors can be similarly applied to what we observe in the case of Lijiang. What these lifestyle migrants pursue is a better quality of life as an escape from China's speedy modernization. When lifestyle migrants elaborate on their imagination of an ideal home, as O'Reilly and Benson (2009, p. 3) argue, narratives articulate ongoing quests to "seek refuge from what they describe as the shallowness, individualism, risk and insecurity of contemporary lifestyles in the perceived authenticity of meaningful places." My analysis concurs with this argument, since the logic of speed can make such a journey of seeking refuge more meaningful in China than in other geographical contexts. Rather than simply being construed as a defense against modernity, the making of home in Lijiang starts with an escape from the metropolis – the symbol of hypermodernity bereft of homely life. As Brickell (2014, p. 270) notes, home unmaking has the potential to "be simultaneously liberating and disempowering in different ways." By migrating to Lijiang, our respondents fulfill the potential of liberation.

While the reasons to migrate to and stay in Lijiang are much more complex than what I analyze here, the respondents share some common features. They cite noneconomic motives as primary reasons for their migration, and undertake lifestyles different from what they knew before, thus downshifting from a hectic life in metropolitan areas to a relaxed lifestyle in Lijiang. They renegotiate the work–life balance so that they can enjoy a good quality of life, though the way of renegotiation varies. For Mei, the decision to leave Shanghai for Lijiang allows him to stay away from high-pressure work to such an extent that he can have more time to drink tea and alcohol; for Xiao, the need to escape the drudge of city living was strong and explicit because his overloaded schedule in Xi'an almost crippled his physical and mental health. Lijiang enabled Deheng and Lily to recover from exhaustion and collapse. Simply put, their downshifting connotes a mentality of opposition to the dictates of an overloaded and speedy schedule. They enjoy the leisure-oriented touristic milieu, but do not reject the idea of transferring such a milieu into concrete business opportunities in the tourism market.

To these lifestyle migrants, the representation of Lijiang as an ideal home demonstrates their efforts to rescue themselves from the condescension of modernity in China's big cities, but inevitably contributes to the enormous condescension of commodification in Lijiang Old Town. Writing about tourism development in Lijiang Old Town, Su and Teo (2009, p. 92) argue that the Chinese state fully seizes authority in heritage management and "inscribes its values on heritage representations, expecting that the town can generate huge economic and political returns." This argument is echoed in a speech by Cui Maohu (Cui 2017), the then Chief of the Chinese Communist Party Committee in Lijiang, who reiterated that "Lijiang's tourism development depends on attractive natural landscapes and ethnic culture, and on the strict preservation and effective use of Lijiang Old Town, giving rise to a pathway of synthesizing culture and tourism for integrated development." This synthesis embodies a localized endeavor to represent ethnic Naxi culture to Chinese domestic tourists, but speeds up the outflow of native town residents to the new city due to forceful displacement in the case of building the Mu Palace (Su 2011) or rent-driven incentives (Peters 2013). Today, native town residents do not live in the town's core area, but merely dwell in some buffer zone such as Wenming and Wenlin villages on the upper Wuyi street (see Chapter 5). As Su and Teo (2009, p. 166) argue, commodification "trivializes the town's cultural significance and results in the museumization of heritage." This conundrum between commodification and cultural significance highlights the politics of heritage tourism in Lijiang, indicating that heritage is preserved for outside consumers – lifestyle migrants and tourists – not for Naxi town residents.

Making Home in its Material and Lived Aspects

Writing about the complex nature of home, Blunt and Dowling (2006, p. 22) argue that "the material form of home is dependent on what home is imagined to be, and imaginaries of home are influenced by the physical forms of dwelling." Following Blunt and Dowling (2006), we can see that homemaking becomes a process of creating and practicing forms of dwelling and belonging, in which people invest material and social elements into a place called home, with regular activities and furnishings, to fashion and familiarize daily life (Baxter and Brickell 2014). This thinking has implications for understanding lifestyle migrants' efforts to make a new home in the chosen destinations.

The ideal home is materially reflected in the spaces of business and dwelling. For instance, Yanzi (female, from Beijing) notes that her guesthouse is a two-story building. When she renovated it, she decided to maintain the first floor as a spacious and comfortable tearoom so that every customer is welcome to drink tea and chat with other customers. She also spends much time in the tearoom for self-contemplation or socializing with tourists. Although she understands that it is impossible to make much money from a four-bedroom guesthouse, she feels quite happy and relaxed in the space she provides. As she explains, "I treat this guesthouse as my home from the heart." In a different case, Lily runs a guesthouse with her husband. When renovating the house, the couple decided to make the courtyard the center of spatial organization and placed flowers, benches, and a tea table in it. Lily has even installed a swing in the courtyard's corner. To her, the swing is her favorite element of enjoyment and relaxation. She notes, "Sometimes I sit on the swing just for loafing. I don't think anything. The swing is large enough to allow me to lie down. I close my eyes and empty myself. It is a good feeling."

Homemaking can go beyond domestic settings to entail a place-based sense of attachment to Lijiang Old Town. Blunt and Sheringham (2019) argue that the boundary between a domestic space of living and an urban space of dwelling is blurred, giving rise to the integrated dynamics of home-city geographies. To them, these geographies involve a simultaneous process – the domestication of the urban and the urbanization of the domestic – through which we can understand the entangled dynamics of city and home. When I ask my respondents about their idea of home, they rarely talk about structural elements such as household registration and property ownership, but focus more on their love for and even indulgence in Lijiang's natural landscapes, including weather, water, and air. For example, Yuhua, a Hong Kong citizen in her thirties, visited Lijiang as a tourist in 2004. Her love for Lijiang and her struggle to leave it have become entwined throughout her postmigration period:

> Lijiang has become an essential part of my life. Definitely! I am so indulged in this kind of lifestyle. It is impossible for me to go back to Hong Kong and live an urban life.... I cannot find a place with such a wonderful living condition. Look at its environment, so beautiful and so natural! The cost of living is not high. No place in China can be compared to Lijiang.

Similarly, Zhaoyi (male, from Beijing) explains his love for Lijiang:

> I really love Lijiang. First of all, its natural environment is very excellent. Another major reason is its pace of life. In the past, my life was heavy with busyness and tiredness. In Lijiang, life is way much easier and slower. I remember that in the first few months after I came to Lijiang, I was still used to finishing things as soon as possible. Half a year later, I started to slow down and become idle. Today's things can be rescheduled to tomorrow, or even the day after tomorrow. I know how to ease up on myself. The lifestyle in Lijiang is really slow.

For both migrants, the incorporation of love for the natural environment into a sense of home results in strong bonds between themselves and Lijiang. While Yuhua draws upon good living conditions and the low cost of living to exemplify her love for Lijiang, Zhaoyi relies on the natural environment and comfortable lifestyle. In doing so, both articulate a discernible contrast between Lijiang and big cities. In this regard, the materiality of home gains significance through contrasts with their frenetic city life before migration.

In Chinese culture, the unity between human and nature is regarded as essential to maintain wellbeing (Tu 2001) (see Chapter 2). Specifically, attaining harmony between the self and the world is a prerequisite to wellness and thus promotes longevity (Huang and Xu 2018). Lifestyle migrants' embrace of the old town's natural environment should not be understood as part of the exoticizing construction of "other" to support the self. Rather, Chinese traditional culture did not antagonize culture and nature or view this relationship as inherently asymmetrical. In the *Analects*, Confucius notes that "The wise find pleasure in water; the virtuous find pleasure in hills. The wise are active; the virtuous are tranquil."[1] From water and hills, people can cultivate the experience of pleasure and strike a balance between busyness and loafing. Once these ideas about water and hills have been reconstructed as categories of thought to shape the relationship between culture and nature, the predilection for natural environment remains an inherent component of Chinese culture and reminds individuals of their roots in Mother Nature. This emphasis on the interrelationship between culture and nature points to a new dialectical thinking of translocal connections through which an acceptable way of living can be made possible.

Although their daily life is full of possibilities as they encounter tourists in the course of tourism business, many lifestyle migrants actually engage in a simple and easy, sometimes seemingly dull, lifestyle. This engagement is a conscious choice. Jiajia's (male, from Dalian in Liaoning province) proclamation that "everyday I do only three things when I have time: listen to music, drink tea, and enjoy sunlight" should be placed in contrast to his past life experience of feverish haste. A similar example is recounted by Liu, who experiences simplicity in his daily routine as a form of meditation. Liu (male, from Guangzhou) had a relatively successful career in selling building material in Guangdong province. Nevertheless, business-driven socializing crippled his body and placed him on the brink of depression. One day he drank too much with clients and became completely intoxicated. Near death, he was rushed by ambulance to the emergency room for a life-saving intervention. On the verge of collapse, he quit his job in 2012 and spent one year traveling in China. After visiting various destinations, he decided to settle down in Lijiang and operate a guesthouse for recovery and fun. To Liu, intentional simplicity is seen as being useful for physical and mental health:

> My life is very simple. I walk my dog at 7:00 a.m. for half an hour and spend one hour doing some exercise. Then I read books, drink tea, and listen to music. In the

> afternoon, my main activities include practicing calligraphy, and playing chess or a bass drum. In the evening I go out to drop by some teahouses and try some good tea. That's a whole day of my life. It is very relaxed and comfortable, isn't it? It is a means of self-cultivation.

Liu's primary daily activities seem simple and even dull, but help him to maintain inner calm. In Lijiang, he has no need to engage in heavy drinking or business problems, but in self-cultivation and meaningful recovery. In part, this theme of simplicity is made possible because Liu and many other lifestyle migrants consciously create a relaxed and comfortable milieu, which is different from hectic city life, for themselves and for tourists as well. As he says: "People emphasize Lijiang is a place for them to escape. Here people can escape from complicated reality." This narrative links into Liu's practices of a simple life to maintain a different lifestyle from what he had before, giving new meanings to his postmigration stay, though he struggles to categorize his choice as a passive escape or an active reclusion.

Simplicity remains a key characteristics of Lijiang lifestyle. For instance, Mei stresses how simple his life is: "Oftentimes I stay alone in the courtyard, drinking tea and enjoying sunlight, for a whole day. I might think about many things, but nothing is important." In a slightly different way, Yang (male, from Beijing), a bookstore owner who migrated to Lijiang in 2011, talks about his mundane activities:

> In the morning, I carry a basket and walk through the old town to buy vegetables in the fresh market nearby. This is an authentic life, replete with autonomy and ease. Why should we wear masks in our everyday life? In Lijiang, you can take down these masks and become your true self.

For both of these migrants, the incorporation of simplicity and ease into their new lifestyle results in the routine of daily activities, an essential component of their ideal home. While Mei draws on tea drinking alone or with tourists to facilitate his reflection, Yang underscores daily shopping for vegetables and fruits to enrich his experience of life. In doing so, both establish a classic link between a good life and an alternative lifestyle. These daily activities, which seem ordinary and uninspiring, point to the essence of *being* in relation to simplicity and relaxation so that they can (re)discover their true self.

Another dimension of a good life is a slow and natural sense of time. Before moving to Lijiang, lifestyle migrants had to cope with frantic haste and high stress. After migration, they can slow down their pace of life without feeling driven by the clock. They reiterate that they have no need to set an alarm in the morning as they enjoy sound sleep. Liu states: "I wake up naturally, have time to drink tea and chat with other people, and am able to do what I love to do." Xiao notes that "different from nine-to-five daily routines in big cities, I get up when I naturally wake up." For both Liu and Xiao, waking naturally is an experience of the good life which they did not have in the past, and shows how thoroughly relaxed they are in Lijiang. This experience was commonplace to our respondents, highlighting the way in which migrants are relieved of overwhelming anxiety and stress, so that they can enjoy cyclical time. While some authors have emphasized that material landscapes are central to the remaking of home for migrants (Boccagni 2017), I emphasize that mundane activities such as waking up naturally are also important for an experience of ideal home to counteract the effects of mounting stress and constant busyness. Simply put, the culture of speed has converted the enjoyment of cyclic time such as waking up naturally into a luxury, particularly when what Crary (2013) calls 24/7 capitalism marks the end of sleep. In Lijiang, lifestyle migrants such as Liu and Xiao prioritize meaning over money so that they can regain such an enjoyment.

A similar indicator of temporal autonomy is a slow pace of living. Huang (male, from Shanghai), a guesthouse owner who was deeply fascinated with the lifestyle in Lijiang, gave preference to the lack of clock time when mentioning how he organized his everyday life:

> Daily life in Lijiang is not special. For businesspersons in Lijiang, we have much autonomy in time management. Very relaxed! We don't have a fixed schedule. No one forces us to get up at certain moment or meet someone on time. Our life is not disciplined by the clock.

Though he understands that it is necessary to make adequate money from the tourism industry to sustain his lifestyle in Lijiang, Huang maintains that earning money is not his priority. Similarly, Hua (male, from Zhangzhou in Fujian Province), a teashop owner who has been in Lijiang for over ten years, explains his sense of time: "I look at the sun's position to organize my daily life. In Lijiang, there is no conception of time. If you ask me which day is today, I have no idea. If you ask me which month, I probably know." Huang and Hua's understanding of time resonates with other respondents who have extricated themselves from the rigid control of clock time and reverted to a free-forum lifestyle. They reject the rule of the clock and discard the "restless urgency to consume time purposively" (Thompson 1967, p. 39).

The temporal aspects of slowness and simplicity are related to Lefebvre's emphasis on the value of cyclic time to the practices of everyday life. According to Lefebvre (2002), everyday life lies at the (dis)connection between the accumulative and the nonaccumulative. On the one hand, everyday life ought to submit to the demands of accumulation, through the discipline of time and space, in order to maximize economic return. If the temporality of the nonaccumulative is embedded in cyclical time, then the accumulation process is characterized by the linear time that results from "the monotony of actions and of movements, imposed structures" around the clock (Lefebvre 2004, p. 8). On the other hand, individuals attempt to distance their everyday life from the process, or become dissociated "in the giddy heights of specialization and technology" (Lefebvre 2002, p. 335). One way to achieve this dissociation is to remain linked to cyclical time that, according to Lefebvre, originates in the cosmic, in nature, and becomes rooted in lived experience. This link is noted by Fei Xiaotong who described how Chinese farmers traditionally revolved their daily life around sunrise and sunset and complied with seasons. In Lijiang, lifestyle migrants make efforts to retrieve cyclical time.

As I explained in the case of Mahua at the beginning of Chapter 1, the making of an ideal home can become attractive to tourists who come from China's big cities and desire an experience of homely life, no matter how transient this experience may be. The synthesis of ideal lifestyle and tourist business can be found in the case of Yunhui's (female, from Shenyang) guesthouse. In her forties, Yunhui quit her job in a large state-owned enterprise and migrated to Lijiang in 2010. Then she chose to run a guesthouse so that she could have a place to implement her cultural value of a simple life and share it with tourists. As she explains,

> I prefer to live a simple life. That's why my guesthouse looks simple and cozy. I want to create a homely atmosphere for my customers. I migrated to Lijiang because of my passion with the lifestyle. I hope my customers have a sense of home and thus feel quite comfortable during their stay.

She emphasized that her guesthouse, like many other in the old town, is different from standardized star hotels because she can create a unique living condition for tourists to interact with each other and make friends (Figure 3.1). Similarly, Liu highlights the theme of meditation and simplicity in his guesthouse, as many of his customers come to Lijiang not for excitement and exotic difference, but for thorough relaxation. This theme, according to Liu, has attracted many repeat tourists, so that he can share his thoughts with tourists and sustain his tourism business as well:

> Many customers in my guesthouse stay in their room for several days. Only in the evening do they step out to have dinner, drink tea, and chat with other people. They don't want to go anywhere. In fact, the major theme of my guesthouse is quietness and meditation. These customers like the theme. They have no interest in visiting other attractions or taking photos. They prefer to live in a quiet place and stay calm.

These shared values between Liu and his tourist customers mean that the practices of a good life in Lijiang revolve around some basic needs: rest, friendship, and meditation. These cultural meanings of ideal home constructed by lifestyle migrants such as Liu and others highlight how a return to inner heart through meditation helps people to feel at home and enjoy themselves in the face of China's great transformation.

Figure 3.1 The Courtyard in Yunhui's Guesthouse in 2020. **Source:** Used with permission from Huang Qingyan.

These lifestyle migrants seem to avoid being entangled in complicated social and economic processes in which profit is maximized. Rather, they deliberately live a simple life in the town, despite the complexities of the tourism industry. To some extent, their deliberations represent a countermovement against speedy modernity, as they choose to withdraw from the race for wealth and fame in China's metropolitan cities. The making of an ideal home by lifestyle migrants should not be simplified as an eccentric model of China's great transformation, nor can we highlight that their choice entails a political rhetoric to challenge the Chinese state's grand project of modernization. Rather, the new lifestyle in Lijiang gives my respondents an at-home feeling, not because they actually own real property, but because they genuinely love the blend of natural environment and cultural milieu. They have a clear idea of the lifestyle that they want to live in Lijiang, with common keywords in their narratives such as "flexibility and relaxation," "a pleasant way of life with low pressure," and "doing nothing but fun." They also display and experience their sense of style in the assemblage of material elements (music, tea, a swing, flowers, buildings, and streetscape) and practices (walking, enjoying a sunbath, and chatting) that they combine into a lifestyle in their own home. Feeling at home, then, remains appealing to Lijiang's lifestyle migrants because ideal home affirms "a profound desire for social solidarity" (Bhabha 1994, p. 27). To my respondents, the ideal home entails a place of comfortable dwelling as well as a form of social solidarity, in contrast with their frenetic city life before migration.

Lifestyle migration is a conscious form of mobility as migrants want to escape to destinations where they can achieve self-cultivation and a new life (O'Reilly and Benson 2009). The purpose is to cultivate "a recreation, restoration or rediscovery of oneself, of personal potential or of one's 'true' desires" (O'Reilly and Benson 2009, p. 3). To many lifestyle migrants, their sense of escape is not temporary, but everlasting to such an extent that they find it challenging to reintegrate themselves into an overloaded urban life. For instance, Mei even jokes that he has become a waste (feiwu, 废物) after spending five years in Lijiang. He further explains that "If you stay in Lijiang for a very long period, you lose the will to fight. It is definitely true. If now I go back to big cities, I am able to play for fun. But I don't think I can work arduously." Being waste-able is an interesting condition of existence, as it shows that Mei and other respondents alike are capable, financially and technically, but do not want to waste their time striving for materialistic gain. They understand that they might not be motivated to resume the hustle and bustle of urban life in those cities they left behind.

These lifestyle migrants should not be defined as either "human waste" (Bauman 2004) or marginalized rural migrants in Chinese cities (Wang and He 2019). Rather, they possess a certain degree of social agency to choose an alternative lifestyle. Ironically, these migrants' escapes from the overwhelming stresses of city life is largely driven by their own individuality within the context of great transformation in which people can warmly embrace excessive individualism in contemporary China. In other words, not everyone is capable of being waste-able, but some with financial and cultural capital can put into practice their own individuality by consciously choosing a waste-able condition in Lijiang. If being wasted represents a totalizing mode of production in the context of neoliberal capitalism, then engagement with a waste of time by these lifestyle migrants epitomizes a subtle and nonviolent resistance against time discipline. The question of who can get rich first and fast must be supplemented by the question of who can waste time at home. The privilege of wasting time, as joked by some lifestyle migrants in Lijiang Old Town, comes from a conscious choice to demodernize their life in relation to an alternative temporality other than time discipline and space efficiency. Furthermore, this privilege draws on their ability to make a profit from Lijiang's tourism industry by selling experiences to those tourists who would like to enjoy a moment of wasting time. Nevertheless, as I will discuss in Chapter 6, this waste requires the imposition of time discipline upon service workers who rush to provide necessary domestic service to keep the whole industry running.

I do not suggest that this group of lifestyle migrants in Lijiang is either homogeneous or equal. Instead, I seek to understand the alternative lifestyle that these respondents attempt to live, in comparison with the bustling lifestyle in which they were once implicated, not for social distinction in relation to other people in big cities, but for dwelling comfortably on their own terms. Nevertheless, this form of dwelling is not bereft of tensions, but becomes embedded in a blend of belonging and alienation.

Unmaking Home: The Spatial Politics of Belonging and Alienation

The lifestyle migrants who I interviewed have dwelt in Lijiang Old Town for years, but have chosen to not integrate into Lijiang's local society. They want to maintain their preferred lifestyle, and only reluctantly get involved in local affairs. When asked whether they have made friends with native Lijiang residents, all respondents proclaimed little interest in doing so. For them, the negotiation of identity involves the confrontation between being local and being outside. Hence, Mei maintains that "I think I am a local. Tourists also regard me as a local. Nevertheless, I never regard myself as a Lijiang native." When I further ask why he holds such an ambiguous identity, Mei explains that Lijiang economically and socially lags behind big cities, and social services such as public education and medical care cannot be compared to coastal cities. Hence, he emphasizes that "I don't want to regard Lijiang as my permanent home. It is just a place of vacation." For Mei, Lijiang's slow life and natural environment make him feel at home, but its relatively underdeveloped support system runs contrary to his sense of home, giving rise to an ambiguous identity between temporary migrant and long-term tourist. Respondents such as Mei regard themselves as Lijiang locals because of their good knowledge of how to live in and enjoy the place. The distance between migrants and natives alludes to a mixture of cultural and ethnic differences in Lijiang. The claims to and knowledge of how to live in Lijiang become a source of belonging. Through these claims, the migrants aspire to a sense of place that is intimately embedded in their experience of ideal home. Furthermore, through possessing local knowledge and appropriating the lifestyles that they favor, these migrants can detach themselves from native residents and thus maintain their identity without conforming to any pressure from local society. This social detachment, as Zhu (2018) explains, is attributed to migrants' efforts to retreat into a self-made oasis, and thus enjoy an exclusive living milieu.

External Pressure for Home Unmaking

Two specific situations confront lifestyle migrants in Lijiang. The first situation comes from the tension between migrants and native residents, which casts a real shadow upon my respondents' residence in Lijiang. The so-called local protectionism, directly and indirectly supported by Lijiang's local government, aims to secure economic entitlement and property ownership for Lijiang natives (Su and Chen 2017). In Lijiang Old Town, about 60% of properties were held by private individuals and 40% by local government agencies (Su and Teo 2009). The former are defined as private properties (私房, sifang), while the latter as public properties (公房, gongfang). Before Lijiang's tourism industry started to boom in the late 1990s and early 2000s, very few domestic migrant businesspersons were interested in buying properties from private owners. Only in 2003 and 2004 a few private individual owners succeeded in selling their properties to migrants for cash. In late 2004, the Lijiang Heritage Authority began to disallow the sale of properties between local owners and migrant businesspersons for the sake of preserving the town's heritage value (Su and Teo 2009). Due to Lijiang's booming tourism industry, the rental price for property in the town kept rising, and local owners lease their properties to migrants for steady economic return. Another policy was to discourage migrant businesspersons from obtaining business permits, but town residents found it relatively easy to apply for business permits. Hence, a private property owner leases two things to migrants: the right to use the property and the business permit to run it as a guesthouse or a souvenir shop. These arrangements stipulated by the Lijiang Heritage Authority have been widely criticized by migrants, but well serve the interest of town residents.

Preservation-related arrangements place migrant businesspersons in a very vulnerable position in comparison with property owners, and can easily generate tensions between the two sides. Making concessions is a common strategy for migrants use to handle conflicts with property owners because these migrants regard themselves as outsiders and lack the necessary social capital to deal with owners who persistently press for increased property rents or to nullify the lease. When neither side can reconcile the conflict, they file a lawsuit, but the local court is constantly alleged to be in favor of Lijiang natives. Here Yunhui stresses the stark divide between migrants and natives:

> We cannot be categorized as Lijiang *ren* (丽江人), or Lijiang native. We simply reside in Lijiang and can be regarded as residents in Lijiang. Many migrants identify themselves as new Lijiang locals, but natives do not acknowledge this identity. These natives always criticize migrants for polluting local culture or over-commodifying the town.

Interviewees such as Yunhui express an insidious aspect of tourism business between migrants and natives. As a result, migrants express a strong love for Lijiang, but they do not regard themselves as natives in Lijiang. They create a sense of community among themselves and become alienated from Lijiang natives. To them, this migrant community is what they can call "local society." Lifestyle migrants' belonging to an invented migration community illustrates how they make efforts to redefine self-identity in relation to cultural values that underpin their relocation. This understanding of belonging and alienation serves as an alert to the power relations enacted and resisted through the material manifestations of ideal home.

Another constraint is the highly competitive tourism market in Lijiang Old Town. Although the town cultivated more than ten million tourist arrivals per year in the 2010s, the tourism market has not been as rosy as many expected. The commitment to lifestyle is generally subject to the coercive law of competition. Beyond initial costs for license transfer and decor, migrant businesspersons need to handle other costs related to property rent, labor, and operations. Today, the starting investment for a tourism business easily places most lifestyle migrants who are not super-rich into a financial plight. For instance, Liu observes how he feels stuck in Lijiang between a favorite lifestyle and a tough tourism market:

> I came to Lijiang for relaxation, for freedom from any burden. After staying for five years, I become overburdened due to the precarious nature of my business. I have invested so much money and made so many efforts to maintain this guesthouse that I cannot easily give it up. I am not really free, but rather have to care about some things. I am stuck in Lijiang.

For Liu, to stay or not to stay in Lijiang is a question that really matters. In particular, he understands the hardship of maintaining his relaxed lifestyle in Lijiang's competitive tourism market. In another case, Ming (male, from Harbin, migrated in 2015) followed his doctor's advice to stay in Lijiang for recuperation. But he did not want to hang around for nothing, so he took over a guesthouse. He explains:

> Easy money? Are you kidding me? We all know how competitive the guesthouse market is! I invested CNY 400,000 (~US$ 59,000) in the guesthouse and felt worried about economic return. It is not easy to run it. I heard that one third of the guesthouses in the town were closed down [in 2016]. When business is bad, I feel mentally exhausted; when business is good, I feel bodily exhausted. Life is not easy, though Lijiang's lifestyle is desirable!

The trouble that Ming encounters is not unique. For those who lack adequate financial capital, their efforts to make a new home in Lijiang are constantly mitigated by the competitive tourism market. Once they cannot sustain their business, they must transfer it to others and leave the town for a new life journey. Some choose to return to their hometown, some go to big cities for new fortune, and yet others search out a different destination with a similar lifestyle and lower living costs.

While the first situation highlights the nature of migration in relation to local natives in Lijiang, the second specifies another role that our respondents play: lifestyle entrepreneurs in a highly competitive tourism market. Similar to lifestyle entrepreneurs in other cases, our respondents run tourism-related businesses and maintain sufficient income to ensure that the business offers them and their family a satisfactory level of funds, enabling them to enjoy their favorite lifestyle (Hall, Lundmark, and Zhang 2021). Ateljevic and Doorne (2000, p. 388) argue that lifestyle entrepreneurs reject an overtly profit-driven orientation in order to "preserve both their quality of life in their socioenvironmental contexts and their 'niche' market position catering to travelers similarly seeking out alternative paradigms and ideological values." These scholars emphasize that the pursuit of certain lifestyles has assumed a symbolic status for those who migrate to tourist destinations and engage in the tourism business. Missing in these insights on lifestyle entrepreneurs is the tension between sustaining ideal lifestyles and running small tourism firms at risk. My analysis of home unmaking in Lijiang sheds light on this tension derived from the coercive law of competition in the face of increasing pressure on economic return in the tourism market. Homemaking is a cultural construction, but due to pressure, this construction is porous and fragile.

Internal Struggle between Here and There

The struggle between Lijiang and elsewhere has been internalized by lifestyle migrants in Lijiang. Regardless of their identities, as migrants or entrepreneurs, my respondents speak of the peaceful and simple way of life that they enjoy very much, but emphasize growing feelings of being uprooted by and disconnected from the local society. Maintaining the possibility of return to big cities can partly be explained as strategic, as it means that they still have the option of resuming the frantic work–life trajectory once they are tired of or fail to sustain their lifestyle in Lijiang. As many respondents reiterate during interviews, "the precondition of my lifestyle is a solid financial foundation. No money, no lifestyle." It seems that a farewell to Lijiang is a daunting task, but the option of departure is always on their radar. As Baxter and Brickell (2014, p. 138) point out, home unmaking is more or less driven by the spatial politics of belonging and alienation from "discrimination, a lack of recognition, and the choice to selectively attend to normative spaces of home and exclude those that do not fit an ideal model." This spatial politics of home is similarly reflected in our respondents' efforts to maintain a "local" community without Lijiang natives. Under pressure to make a profit in the tourism industry in Lijiang, some of them do not find a "fit" in this community and choose to leave.

Indeed, lifestyle migrants' attachment to Lijiang is not unequivocal when they talk about the variegated meanings of home in relation to their place of origin. To Hua (male, a teashop owner, from Zhangzhou), who has lived in Lijiang for thirteen years, his sense of belonging always has been geared towards his hometown in Fujian province. As he emphasizes, "I have a house in Fujian.

In Lijiang, I think I am just a part-time local. It is impossible to develop a sense of belonging. I have friends and hometown fellows in Lijiang, but I always feel strange in a familiar place." His thirteen-year-long stay in Lijiang has not eased his lingering sense of rootlessness, even though his whole nuclear family also has settled down in Lijiang. Hua does not want to completely leave Lijiang for his hometown in Fujian. Instead, he has short visits to Fujian and then rushes back to Lijiang. He lives in an in-between condition: "In June 2016, I returned to my hometown. After staying for five days, I felt very uncomfortable because of the humid and hot weather. I then returned to Lijiang and felt way much better." Although he seems to express feelings of in-between – feeling at home in and uprooted from Lijiang – Hua's choice of unhomely life hints at something deeper. It suggests a constant negotiation for feeling at home between Lijiang and his hometown. He spoke of a peaceful and simple way of life that he enjoys very much, but emphasizes growing feelings of being uprooted and unrelated in a place where he has established his tea business, raises his children, and is exposed to tourist customers.

This kind of ambivalent feeling is similarly expressed by Jiajie (male, from Nanjing). He dislikes Lijiang's lack of nightlife and good cuisine, but seems to have become used to Lijiang's lifestyle:

> I feel bored to death after staying in Lijiang for two or three months. Then I travel to other places and return to my hometown. Nevertheless, once I leave Lijiang for one month, I miss it and need to travel back to Lijiang. My life repeats the process again and again.

Here Jiajie expresses a prevalent feeling of in-between – feeling at home in Lijiang and also disconnecting from the place. This in-between condition spells out an unhomely life and suggests a constant negotiation for a sense of belonging between Lijiang and elsewhere. Having said this, I emphasize that Lijiang's reality is an economically and socially disadvantaged place compared to China's big cities such as Beijing and Shanghai. Lijiang's backwardness clearly upsets Jiajie, but does not cause much of a stir among many other respondents I interviewed since they realize that they cannot have both money and meaning at the same time as far as the deal of modernity is concerned.

One main reason for these feelings of rootlessness or disconnection is that lifestyle migrants do not grow up or build social ties with others from childhood in Lijiang. Yuhua (female, a souvenir shop owner, from Hong Kong) visited Lijiang as early as 2004 and really loved the town. Without much hesitation, she decided to leave Hong Kong and stay in Lijiang. To her, Hong Kong was filled with high pressure and a hectic pace. At the time of our interview, she had married another lifestyle migrant, raised their two children in Lijiang, and had no plan to leave. Regarding what serves as a place to call home, Yuhua sees the difference between rooted home and sensed home by lamenting that:

> Frankly speaking, Lijiang is not a place where you grow up. In such a place, you can have a group of friends who grow up together with you. So here I don't have a sense of rootedness. I do make many friends in Lijiang, but feel it is not my place. If I had grown up here, I might have a different feeling. I cannot develop this sort of feeling in Lijiang. You cannot let time to go back to childhood.

Yuhua recognizes that Lijiang has become an essential part of her life, because of her family as well her personal preference. To her, Hong Kong is the place where she grew up and had a lot of lifelong friends. Meeting these friends to renew friendship is a major incentive for her to revisit Hong Kong, a place that she calls "hometown," but the hometown is unbearable after a several-day stay, so she desires to return to Lijiang. As Yuhua explains, "So many people walk along streets in Hong Kong. Very crowded! The only pleasant thing I do in Hong Kong is to meet my old friends. I don't want to leave them, but I know I need to return to my own life in Lijiang."

Indeed, Lijiang becomes a place for Yuhua to return to, and in this sense, it can be defined as a home for sure. If Hong Kong can be regarded as a rooted home that exists in Yuhua's nostalgic sentiment, Lijiang becomes a sensed home that gradually embeds her in a setting where she feels comfortable and able to enjoy. The meaning of home is marked by an emotional attachment to a place, a person, or a community, and this attachment keeps ringing the bell of going home. Furthermore, home acts as a bridge to (dis)connect the self and the world. As Tucker (1994, p. 184) explains,

> Home is where we could or can be ourselves, feel at ease, secure, able to express freely and fully, whether we have actually been there or not. Home is the reflection of our subjectivity in the world. Home is the environment that allows us to fulfil our unique selves through interaction with the world. Home is the environment that allows us to be ourselves, allows us to be homely.

These activities of "feeling at ease" and "fulfilling our unique selves" in the making of home and, by implication, the relationship between the self and the world, are all experienced by my respondents in Lijiang. For Yuhua and others, there is a need to secure spatial and temporal autonomy in a rapidly changing society where individuals are integrated into a restless process of capital accumulation and subject to the ubiquitous existence of political control. Meanwhile, they are torn apart by an in-between condition – Lijiang and elsewhere – so that they can feel at home in Lijiang, but this feeling is fleeting and unrooted. They engage in a life journey in search of home, bridging the gap between the ideal home where they dream of a fulfilled self and the actual home, which is besieged by the concurrent logic of profit making and political discipline.

The simple question "where is a place called home" becomes a difficult one for many lifestyle migrants in Lijiang to address. This in-between lifestyle is best narrated by Yang (male, a bookshop bar operator, from Beijing) who articulates that "Lijiang is not a permanent site for residence. I just live here for a very long time" and by Limi (female, a souvenir shop owner, from Guangzhou) who laments that "If I were born in Lijiang, I think I could feel pride in it. To be frank, I do not belong to Lijiang. I just temporarily live in Lijiang." Meanwhile, Xu (female, a guesthouse operator, from Harbin) really loves Lijiang Old Town for its natural environment and simple lifestyle, but emphasizes: "There is an old saying: east or west, home is best. Home is the best place no matter how long I stay in Lijiang. My hometown is Harbin. I miss my hometown so much."

Since 2008, the Lijiang city government has realized the value of lifestyle migrants to create an attractive milieu in the tourism market. Hence, the government launched a campaign of "Being new Lijiang people" (xin Lijiang ren, 新丽江人). The goal is to persuade these migrants to settle down in Lijiang and to become socially embedded with local people. Despite the strong push from the Lijiang municipal government, lifestyle migrants do not subscribe to such a political agenda that aims to place them in a conundrum of permanent settlement. Rather, they strategically become involved in the tourism industry, together with the cultural and management resources that they had already acquired before migration, to ensure their enjoyment of their favorite lifestyles, in ways that the local government cannot initiate. Thus, they continue to detach themselves from local affairs.

So home must be regarded as a complex and mobile entity of imagination and practice – an imagination of an ideal place that offers comfort and autonomy and a practice of modern life filled with loafing and profit. To these migrants, Lijiang is not a permanent setting for them to satisfy their desire for a better way of life. Rather, it is just a stop in their individualistic life journey, though they rarely know how long they will stay. They have the privilege to choose their life in Lijiang and elsewhere and to realize their pursuit of relaxation and simplicity within their daily life. Once they lose this privilege, they launch a new journey to try something different or resume a hectic life for money. To them, life should be filled with possibilities, and their migration attempts to fulfill one possibility among many. They become lifestyle hunters in an era of liquid modernity (Bauman 2000). They celebrate the permeability of boundaries between here and there, between one destination and another, enabling their home-searching adventure to be in-between.

Divergence between Busyness and Slowness

Generally, guesthouse owners reported that they relied on online travel agencies (OTAs) and WeChat Friends to promote business and obtain reservations.[2] Due to the market downturn caused by the COVID-19 pandemic since early 2020, tourist arrivals in Lijiang have plummeted from 35.15 million in 2019 to 18.91 million in

2020 (Bureau of Statistics in Lijiang Old Town District 2021). Given this condition, some guesthouse owners began to switch marketing platforms from OTAs to livestream media. Initially, a few made short videos on their own and uploaded to Douyin (抖音, a counterpart of TikTok in China) or Kuaishou (快手), two of the largest self-media platforms in China, to attract customers. More and more business owners in Lijiang embrace self-media and spend much of their time making short videos so that they can promote their business and compete for tourists. For those who have secured tens of thousands of followers and become influencers, they have realized the huge potential of livestream sales. In addition to promoting their own guesthouse, they use livestreaming commerce to sell Lijiang's local products such as tea, silverware, and cosmetics. Video making replaced homemaking as the kernel of Lijiang's hospitality business.

In August 2019, Wang, the guesthouse operator I mentioned at the beginning of this chapter, spent most of his time drinking tea and reading books. He had plenty of time to taste different types of tea and perform the art of tea making and drinking for customers and friends in his guesthouse. Tea drinking can bring benefits such as improving physical health and clearing the mind to reduce stress. The social benefit cannot be underestimated. As Wang mentioned, "Our customers come from different sectors. Around the tea table, everyone is willing to share thoughts and contribute to interesting discussion. Thus, I can learn from their experiences and broaden my horizon." In October 2020, however, Wang had less time to taste tea. Rather, he had to work with his wife Xue to embrace self-media, mostly short videos publicized in Douyin, for guesthouse marketing and livestream sales. Rather than spending unrushed time drinking tea with customers, Wang focused more on new ideas for video making so that his guesthouse could secure a steady inflow of tourists. In the past, guesthouse owners could do business with ease and enjoy a relaxed life. When the market turned bad, they kept busy marketing their business and some even started e-commerce for more profit.

Ironically, guesthouse owners and their teams keep busy making videos to demonstrate a leisurely and slow lifestyle, including afternoon tea, sunbathing in the yard, reading books, dinner parties, and drinking wine or other alcohol. Xue regularly posts short videos to publicize how relaxed and enjoyable her guesthouse is, and the whole team carefully designs videos and tell good stories. It is time consuming and constantly requires new themes and ideas. Thus, Wang emphasizes that the new model of business marketing in the post-COVID-19 period has utterly changed his life:

> Self-media generates profound impacts on our life. For instance, we lose temporal autonomy. In the past, my wife and I could take a trip without a plan, because we had free time and enjoyed a freewheeling lifestyle. We had only one assistant to help us. Now we hire one full-time manager and two assistants. Theoretically, we should have more time to travel elsewhere for several days and drink tea on a sunny

> afternoon. Ever since we started self-media in March 2020, we cannot go anywhere. Self-media has occupied our time and driven us nuts.

In a rapidly changing society, Wang and his wife must make constant adjustments in order to sustain their hospitality business and make money. They came to Lijiang Old Town in 2013 to evade the hustle and bustle of urban life in China's big cities. The COVID-19 pandemic became a turning point, pushing them to embrace self-media to keep the business going. This new direction disrupted the homely lifestyle that they endeavored to maintain for tourists. To compensate the erosion of home, Wang and Xue ask their employees to provide professional services to customers. As Wang admitted, they have no choice, as they do not have as much time to interact with customers as they did before. As her guesthouse's cleaner (A'Jie Li, see Chapter 6) explains to us privately, "[Xue] goes to bed very late, very very late. She seems very exhausted. Livestream sales is a tough business. Sometimes she is too busy to have dinner. She rushes to respond to her customers and confirm purchase orders. How busy she is!"

Like Wang and Xue, many lifestyle migrants in Lijiang Old Town resume a hectic work–life pattern, the one they dreaded and escaped previously. Their reengagement with speed seems to result from their own decision, in order to adjust to China's new business environment, but the primary purpose is the same: the pursuit of wealth and fame. Possibly they realize that they cannot escape any more, since Lijiang Old Town might be their last line of defense to mitigate speedy modernization. Rather than escape from speed, now they choose to embrace it not for someone else's success, but for their own. After all, they are their own bosses who can maximize the use of time and space for economic return.

While Wang and Xue represent a group of guesthouse owners embracing busyness, some others stick to loafing and choose to do business as usual, for two reasons. First, their number of guest rooms is small, generally less than eight, which means that they need not worry about the inflow of tourists. They can rely on OTAs and WeChat Friends to promote their business and receive reservations. Second, they dread the rat race for wealth, the race that once implicated them in the manacles of alienation and homelessness. Seeing what happened to those who engage with self-media, they realize that this new business model can deprive them of temporal autonomy and fill them with dread. Mahua, the guesthouse owner I discussed in Chapter 1, decided not to embrace the incessant race for customers. Regarding her decision, she emphasizes that her original idea of migration to Lijiang was not for money, but for a different lifestyle. She does not want to repeat her Beijing nightmare. Nor does she give up her original idea. As she explains, "big investors start to encroach into the old town, and their priority is profit. The whole business environment is changing. Self-media is one form of adjustment, but I hope to continue my own way."

The divergence among guesthouse owners in the case of making home or making videos reveals the spiral of unhomely life in Lijiang's hospitality business.

The logic of profit making puts pressure upon these owners and pushes them to make a deal between money and meaning. They escape big cities to settle down in the old town and experience an alternative time-space, but in a highly competitive market some actively adjust their business model by resuming a hectic pace for more profit, while others adhere to their original intention and strive for a balance between home and profit. This divergence is a barometer of social and economic change in today's China, since individuals must figure out how to handle the deal of modernity. While the COVID-19 pandemic served as a catalyst for change in Lijiang's hospitality business, self-media brings radical innovations in business operation – video-based promotion, livestream sales, instant orders, and a hectic daily pace – which unleash dynamic forces of social and economic change upon these lifestyle migrants. This is, then, a situated story, one that points to the need to develop a nuanced analysis of homemaking at different stages of life. The changing temporalities that some lifestyle migrants have undertaken in Lijiang demonstrate a necessary adjustment between busyness and loafing so that they can renegotiate the deal of modernity. Some others maintain a usual pace of living and choose not to trade meaning for money. The divergence cannot be explained solely through the theories of political economy – monetary exchange, surplus time, and state capitalism. Rather, this divergence operates through individual agency in a rapidly changing society to negotiate for a sense of home.

Conclusion: The Ambivalence of an Ideal Home

Accounts of homemaking can be gleaned from the works of various theorists of fluid modernity, who argue that it is precisely the in-between flexibility and transience of mobility that create the uncanny condition of belonging and alienation. This condition is evident in the contemporary surge of variegated forms of uncertainty and ephemerality – from the moral critique of social distrust to the quest for inner calm, and from the pursuit of material wealth to the disillusion with modernity. Homemaking is enmeshed in a perpetual battle between order and "the incessant change which is the condition of modernity" (Cresswell 2006, p. 20). As I have emphasized here, an ideal home exemplifies the basic premise of place attachment, outlined by Blunt and Dowling (2006) as a process of engineering and mobilization through imaginations and lived experiences, to pursue a better quality of life. The efforts to make and unmake an ideal home in destinations reflect an endless life journey through which individuals negotiate the dynamic relations between the self and the world.

I acknowledge the ambivalence of the "ideal home" as culturally and spatially situated, and potentially overlapping with feelings of comfort, nostalgia, and even romanticism. I take a broad approach to the notion of the ideal home in order to bridge the pursuit of a better way of life as a subjective experience together with the actual practices of everyday life within certain social and spatial contexts.

Representationally, these lifestyle migrants in Lijiang Old Town tend to view and even idealize their new home, though they fully understand that the reality is far from ideal. Nevertheless, this is their choice, an active choice out of their own heart to say farewell to their hectic life in big cities and to live a homely way of life in Lijiang. This idealization underlies the affirmation of personal resonance with their surroundings for comfort and unity, which requires social and infrastructural support in built environments to help individuals to retrieve the meaning of time and space. This portrayal of an ideal home in Lijiang Old Town reflects a romanticized notion of home, in particular contrast with the hectic urban life of big cities in China, so that these lifestyle migrants can mitigate the uncanny sense of homesickness.

Experientially, lifestyle migrants choose to slow down their daily pace and try not to think too much about economic pressure. They left their previous home in China's big cities and settled down in Lijiang to downshift their life and enjoy idealized lifestyles. They regard themselves as locals in relation to tourists, thanks to their knowledge of and long stay in Lijiang, but as outsiders in relation to Lijiang natives, because of their lack of local roots and their different lifestyle. They create a "localized" community of their own so that they can develop a sense of belonging. Due to social alienation from Lijiang's natives and fierce market competition, some of them face social and economic struggles and even leave Lijiang for another destination. No matter how much they strive to idealize and enjoy a good life in Lijiang, their new home is riven with "feelings of belonging and attachment, alienation and detachment" (Blunt and Dowling 2006, p. 100).

Recently, scholars in China have expressed a strong interest in lifestyle-oriented domestic migration, with a focus on social interactions among lifestyle tourism entrepreneurial migrants (Xu, Ma, and Jiang 2017), living strategies in destinations (Ma and Xu 2016), and social integration structure (Lu, Zhang, and Wang 2017). In the case of Lijiang Old Town, Wang, Bai, and Bao (2021) articulate five dimensions of place integration by lifestyle migrants, including enjoying natural environment, accumulating economic resources, developing social ties, embracing local Naxi culture, and building place-based attachment. In an interesting essay on lifestyle migration as a utopian pursuit, Zhang (2021) argues that these migrants in China are driven by the desire for a better life to escape the ills of modern life in big cities and settle down in peripheral tourist destinations – old towns and rural villages – to run guesthouses, cafés, or other tourism-related business. Lifestyle migration itself has become a popular choice among emerging middle classes in Chinese cities in the pursuit of the natural environment and cultural authenticity and to rejuvenate some lifestyles of premodern China (Zhang 2021). Zhang's observation is particularly valuable, as it points to the lingering cultural tradition of reclusion advocated by Tao Yuanming in ancient China.

The ambivalence of the ideal home is explicit. Even for Tao Yuanming, an ascetic retreat was not what he wanted. His statement that "Back again: after so long caged in that trap, I've returned to occurrence coming of itself,"[3] highlighted a desire for the natural way of life, a desire shared by Lijiang's lifestyle migrants who are aware of their caged daily life and attempt to escape from it. Nevertheless, Tian (2005, p. 101) emphasizes that the natural way of life that Tao advocated represents "a highly idealized state of nature, just like the vision of utopia in Dao de jing."[4] As Tian (2005, p. 101) further explains,

> [Nature] can hardly be sustained for long if completely left to itself; it requires vigilant control. It is, to put it simply, a very unnatural nature, a nature that needs constant artificial care, a nature, like a garden, that has been "naturalized" and made accessible to human comfort and affection.

Thus, Tao Yuanming chose to gaze upon the southern mountain from a distance, but not to stay inside the mountain. "Gazing from a distance" allowed him to maintain ties with local literati and officialdom through scholarly exchange and regular socialization, while "staying inside the mountain" would completely exclude him from the civil world. In other words, Tao Yuanming engaged in a semi-reclusion (banyin, 半隐), between the imperial court, which subjected him to the administrative cage, and the wild southern mountain imbued with hardship and suffering in a purely natural setting.[5] An ascetic reclusion into the nature requires a substantial burden of caring about everyday life, a burden that Tao Yuanming could not easily bear. This burden is not a problem in Lijiang, however, since business operators in the town collaborate with the local government to create a naturalized environment comfortable and safe for themselves and leisure tourists (see Chapter 4) to enjoy the experience of reclusion or retreat with ease. The irony is that the lifestyle of semi-reclusion can be sustained only through tourism businesses and is subjected to the logic of capital accumulation from which these migrants endeavor to escape.

Driven by a conscious ethos of autonomy and comfort, lifestyle migration is often regarded as a privileged mode of mobility. Nevertheless, being privileged does not necessarily mean that lifestyle migrants can escape from this not-so-ideal world and indulge themselves in self-centric living conditions. By analyzing the struggles and tensions that lifestyle migrants endure in Lijiang, this chapter demonstrates that lifestyle migration is not only invested with cultural values and personal significance, but is politically and economically embedded in the context of speedy modernity, political control, and capital accumulation. Downshifting involves a subjective choice of intentional simplicity, but this choice is also subject to the broad structure of political economy and identity politics. With specific respect to Lijiang, a clear tension can be found between migrants and natives, and between the rooted home these migrants left behind and the sensed home in Lijiang.

The ideal home must be regarded as a complex and mobile entity of imagination and experience: an imagination of an ideal place offering a better quality of life, and an experience of alternative lifestyle filled with belonging and alienation.

The process of making and unmaking, through a journey-based analysis of home, is also important to advance a critical geography of home. As Blunt and Varley (2004, p. 3) emphasize, this analysis should focus on thresholds "between memory and nostalgia for the past, everyday life in the present, and future dreams and fears." Furthermore, the juxtaposition of making and unmaking envisions the spatial politics of control and resistance, a contentious process in which "people's domestic lives are rarely fixed or predictable, but rather dynamic and varied" (Baxter and Brickell 2014, p. 135). While a complex matrix of political and economic forces is at work to structure how lifestyle migrants imagine and experience ideal home in destinations, the pursuit of ideal home is explicitly not incidental to the logic of capital accumulation and the ideological axiom of social advancement, but implicit in people's efforts to make and unmake ideal home. The divergence between busyness and slowness among lifestyle migrants reflects variegated responses to the political economic forces in the post-COVID-19 period. Focusing on the ideal home as a site of belonging and alienation generates new possibilities for understanding the spatial politics of home even among those relatively privileged migrants in contemporary China. In Lijiang, this ideal home is not merely for lifestyle migrants, but also becomes highly attractive to some independent tourists who similarly escape social acceleration and time discipline in China's big cities and pursue a sense of home elsewhere, a topic I turn to in the next chapter.

Notes

1 (知者乐水，仁者乐山；知者动，仁者静，《论语.雍也》, translated by James Legg. Source: https://ctext.org/analects/yong-ye/ens
2 Owned by Tencent, WeChat is the largest social media platform in China. At the end of March 2023 the platform had over 1.3 billion monthly active users worldwide and daily over 700 million users read messages in their Friends' circle. Guesthouses owners in Lijiang constantly publicized new promotions and activities via their WeChat accounts.
3 In Chinese, 久在樊笼里，复得返自然, translated by Hinton (2005).
4 Dao de jing (道德经) was written by Lao Zi as a philosophical text of Daoism. Translated as the classic way and virtue, Dao de jing emphasizes naturalness (ziran, 自然) and non-action (wuwei, 无为) as two essential ways in the world.
5 The idea of semi-reclusion is suggested by Weijun Chen, a faculty member at the University of Oregon.

Chapter Four
The Act of Retreat: Tourism, Loafing, and the Consumption of Home

This chapter adopts a temporal perspective to analyze how a group of domestic tourists experience a sense of home through loafing in Lijiang Old Town. Just as space is socially constructed, so is time. As Adam (1998) argues, all social life is timed, or arranged through a time-based order of synchronization and standardization. Adam (1998, p. 202) further emphasizes the need to examine specific formations of "time, temporality, tempo, timing, duration, sequence, and rhythmicity as the mutually implicating structural aspects of time." These formations are lived and experienced via the various meaningful interactions between individuals and their surroundings, made in and through everyday life, to complexify the passage of time not as linear, but as a dynamic process of negotiation and struggle.

Lefebvre (1991b) notes that a dramatic transformation of social time has not only led to the marginalization of cyclic time, but also has altered the structure of temporality towards efficiency and speed. He points out that where the cosmos establishes the natural rhythm through cyclic activities, modern individuals are increasingly alienated from a natural use of time and succumb to the linear mode of time through labor, production, consumption, and leisure. Thus, the formation of social time is no longer in any significant way an expression of human nature, but gives way to what Lefebvre (1991b) calls programmed everydayness. No longer driven by the rhythms of nature, or attachment to the natural passing of time, modern subjectivity has been increasingly implicated in a socially constructed sphere of time discipline, in which a strict temporal order is imposed

Unhomely Life: Modernity, Mobilities, and the Making of Home in China, First Edition. Xiaobo Su.

to orchestrate human activities toward the collective drum of machine time through standardized and predictable precision. While Lefebvre (2002) emphasizes that the development of humanity depends on strategies such as political participation and public self-disclosure, I shift attention to more personal experiences of temporal autonomy for loafing.

The predominance of busyness in contemporary societies has been widely examined in the fields of cultural studies (Sullivan 2008; Wajcman 2015) and sociology (Rosa 2013; Rosa and Scheuerman 2010; Torres 2021). Some recent psychological studies explain that people desire busyness and dread idleness (Hsee, Yang, and Wang 2010). Specifically, busyness is supposed to generate psychological benefits by consolidating effectiveness in task completion (Wilcox et al. 2016) and demonstrating social status (Bellezza, Paharia, and Keinan 2017). While Yang and Hsee (2019) contend that people embrace busyness in order to avoid the unproductive use of time, I present a counterargument that people engage in loafing in order to escape from and resist the culture of busyness. The preoccupation with speed and busyness distracts individuals both from mindful engagement with the world, and from authentic self-reflection, which has resulted in the prevalence of boredom as "an indelible characteristic of modern life" (Chang 2009, p. 39). For many people, tourism has become a form of therapeutic consumption through which they can escape from the pressure of social advancement and hectic schedules and enjoy a temporary retreat.

The focus on the temporality of tourist consumption underpins emerging discussions about how tourists change their pace to search for new experiences and pursue possibilities of autonomy and jouissance (Clancy 2017). The early writings in tourism studies emphasized a temporal difference that made tourism distinct from everyday life through a heightened period of time (MacCannell 1976), a suspension of time and tensions (Daniel 1996), or an experience out of time and place (Turner 1973). These scholars emphasize a fundamental contrast between the ordinary time at home (related to work) and the nonordinary status away from home, which offers valuable references for understanding the passage and use of time between here and there. During this rite of passage, as Turner (1973) argues, tourists enter a liminal phase, "neither here nor there," in which the structuring properties of time – efficiency and busyness – no longer apply so that they can experience an alternative temporality from that of the everyday. While the cult of speed becomes a defining symbol of modernization, the experience and value of alternative temporalities enables tourists to articulate a new form of subjectivity and meaning. Foley (2017, p. 18) argues that the art of wasting time by tourists, driven largely by the slow movement, is to develop "a leisure ethic that values idleness, free time, sociability and unproductive activity."

A retreat from daily routines is necessary to enhance a sense of self and thus enhance subjective wellbeing. As Smith and Kelly (2006) point out, the cult of speed has engulfed Western societies to such an extent that the Slow Movement emerged to counter the cult and encourage a range of mindful practices. Today, the contrast between ordinary time at home and extraordinary time in destinations is reversed, as more and more tourists aspire to experience a different pace of life and escape from hectic daily schedules. In a different geographical setting, they can cultivate a new sense of time by slowing down to unfettered relaxation and socializing with other peers without serious purpose. By doing so, tourists can retreat from busyness and overloaded life to another way, of relaxation and enjoyment, generally a positive spatiotemporal shift with which they can feel content.

In Lijiang, the temporality of tourism can be ordinary or extraordinary, depending on a range of factors, including the types of tourists and the purpose of their trip. Tourists in Lijiang face less obligations concerning what to do with their time. Some tourists keep busy in pursuing extraordinariness, including one-night romances, exoticized Naxi ethnic culture, or drunkenness (Su and Teo 2009). Some establish a different rhythm by engaging in loafing through meaningful, but tenuous, connections to the place. Instead of seeking to treat any particular ailment or illness, these tourists share a common interest in loafing so that they can stay away from the stress and anxiety that they must bear in everyday settings. I study the second group of tourists who experience a leisurely rhythm in stark contrast to busyness and time discipline. These tourists conduct activities that would be impossible in their regular settings and thus enjoy a new mode of existence that embraces an alternative time in a different space.

From a temporal perspective, this chapter analyzes how tourists break away from the cult of speed to experience loafing in Lijiang and cultivate a sense of home. These experiences can be intrapersonal via solitude, or interpersonal via socialization. In addition, it shows that slowness and busyness are not antagonistic, but join together to reflect how tourists escape from and conform to the logic of time discipline. Loafing articulates a valuable form of tourist experience that is shaped by individualistic agency to mitigate speedy modernity. Thus, tourism for loafing exemplifies what de Certeau (1984) calls "the tactics of consumption" through which tourists can suspend daily patterns of time discipline and escape economic rationality. The politics of everyday life developed by de Certeau can provoke much-needed thoughts on loafing as a mundane practice that supplements yet resists the cult of speed.

Solitude and a Natural Way of Living

How to experience loafing is a subjective matter. Some tourists simply choose to arrange their daily schedule without external order. One key tactic is to wake up naturally. Within the context of 24/7 capitalism, Crary (2013) emphasizes that time should be kept in motion in order to sustain production and consumption. Nevertheless, sleep becomes "an uncompromising interruption of the theft of time from us by capitalism," as it is embedded in the natural rhythms of life oscillating between solar light and darkness, activity and rest, and work and recuperation (Crary 2013, p. 10). Like lifestyle migrants (Chapter 3), retreat tourists cherish waking up naturally and enjoy sound sleep without worrying about ringing phones and buzzing alarm clocks. To them, this form of rest affirms the necessity of temporal autonomy.

Through waking up naturally, these retreat tourists realize the moment of loafing. Yue, a female tourist from Xi'an, started to travel to Lijiang in 2010, and at the time of interview, she was on her fifth visit. In Xi'an, Yue's hectic daily rhythms make her feel depressed, but in Lijiang her rhythm changes. She explains, "In Lijiang, I wake up naturally. No need to worry about a clock, or going to work, or thinking about job performance. To me, this is a sort of surreal life." We need to place waking up naturally in relation to cyclic time in the broad context of political economy. At the dawn of the Industrial Revolution, Europe started to practice time discipline under the guise of puritanism (Thompson 1967). Though fraught with conflict, the transition from cyclic time to linear time placed a strain on newly emerging factory workers and forced them to use time productively (Foucault 1979). While cyclic time is associated with the rhythms of nature, linear time is acquired through technology (e.g., the clock) to promote production (Lefebvre 2002). Specifically, cyclic time originates in the cosmos, in nature, and becomes rooted in lived experience, while linear time is a social and economic construction resulting from "the monotony of actions and of movements, imposed structures" around the clock (Lefebvre 2004, p. 8). The temporality of capitalism is the principle of nonidleness, and those who refuse to purposively use time or simply waste time would receive severe punishment. Accordingly, the value of busyness has become a necessary component of "the moral fabric" of modern capitalist societies (Ehn and Löfgren 2010). Time discipline has dramatic effects on the health and safety of people, through both hectic work hours and the commodification of work-free time. Even sleep has become a barrier to capital accumulation and sleepless life is promoted to hasten the exhaustion of time, generating health hazards for individuals and placing their subjective well-being in jeopardy (Crary 2013). In Lijiang, tourists are essentially free from their social and economic obligations. Within certain circumstances, they can stay away from the clock and wake up naturally.

In addition, Yue prefers to sit on a bench alone to watch other tourists come and go. For Yue, "siting alongside the street is a way of loafing." Yue feels as though daily pressures and annoyance disappear during her stay in Lijiang, adding to her sense of loafing, a sense characterized by her as "surreal." Relieved of pressure and busyness, Yue can immerse herself in an idle rhythm in Lijiang, though the town itself is extremely bustling and noisy, as tens of thousands of tourists joggle in the streets and alleys during the daytime and at night. The trips between Xi'an and Lijiang herald a switch between busyness and slowness, a switch between struggle and happiness. As Yue explains,

> Whenever I visit Lijiang, I feel very jubilant, because I needn't think about various trivialities in daily life.... From surreal life in Lijiang to real life in Xi'an, an adjustment is needed. I remember that after I returned from Lijiang the first time, I disliked Xi'an so much that I couldn't fall asleep. I cannot help but miss waking up naturally in Lijiang. Then in the morning the damned clock woke me up. It took me about one week to recover from my trip and get used to Xi'an again. Now it takes me two to three days to adjust.

It is hard to articulate how much Yue benefits from loafing in Lijiang, but her repetitive visits provide her with a set of alternative temporalities that allow her to release pent-up anxieties acquired in Xi'an. For Yue, loafing in Lijiang and busyness in Xi'an enter into "perpetual interaction and are even relative to one another" (Lefebvre 2004, p. 90). This interaction characterizes the rhythmic variations of people like Yue and others who enjoy loafing in Lijiang and bear the burdens of busyness in their regular abode.

In another story, Zhou, a female tourist from Wuhan, has visited Lijiang every year since 2014, and several times in one year. Being self-employed, Zhou can arrange her frequency of travel. Her longest period of stay in the town was about two months. At the end of each stay, she feels bored of daily routines and wants to do something back in Wuhan. After a while, she again feels burnout, and then returns to Lijiang for a break. As she observes, some people love big cities and want excitement and busyness, but she leans toward a quiet and simple way of life. Regarding how she spends her time in Lijiang, Zhou comments that she wakes naturally around noon and has free time to do some leisurely activities until dinner. After dinner, she either jostles in the town for shopping or watches movies in the guesthouse where she frequently stays. Here is a conversation between me and Zhou:

Author: What do you do in Lijiang?

Zhou: I rarely go to any attraction sites. My purpose is to enjoy slow life.

Author: What kind of slow life?

Zhou: Just like I sit in this café to drink afternoon tea and enjoy sunlight. I enjoy loafing. To me, loafing means to empty myself, listen to music, read books, and even write some postcards. I prefer to stay alone, just myself. When I am alone, I can reflect on some things and quietly enjoy this slow pace in Lijiang. I don't want to be bothered.

Author: What is the best way of loafing in the town?

Zhou: On the deck overlooking the old town in the guesthouse where I live, I sit on a soft bench to enjoy sunlight and gentle wind. I can either read books or view the whole town. That is what I love to do. The guesthouse is decorated with fresh flowers everywhere. I like this sort of natural beauty. Actually, I don't enjoy daily life in big cities. People are suspicious and jealous of each other. I feel quite exhausted. Once

> people encounter troubles in their personal life or job, many choose to visit Lijiang for an escape. I think my traveling is to escape Wuhan for the purpose of a different experience in Lijiang.

Navigating everyday life in Wuhan is arduous to Zhou, particularly due to high pressure, ubiquitous competition, and hectic pace, giving rise to negative consequences such as exhaustion, alienation, and strangeness. On many occasions she would struggle to figure out the meanings of life. In praise of slowness, Milan Kundera (1996, p. 39) has written, "The degree of slowness is directly proportional to the intensity of memory; the degree of speed is directly proportional to the intensity of forgetting." In contemporary China, the cult of speed plunges individuals into a condition of amnesia. To counter such a condition, Zhou frequently travels from Wuhan to Lijiang. By enjoying well-maintained landscapes and engaging in self-reflection in Lijiang, she can regain inner calm and retrieve a sense of home. As she explained in her moments of loafing, Zhou's experience points to spatial and temporal autonomy. Time, in this context, is not for productive economic work, but entails modalities of relaxation and slowness. While such forms of relaxation are typically seen as noneconomic or nonutilitarian, they bring to retreat tourists such as Zhou a sense of home, as I will discuss later.

Loafing is cultivated through an autonomous use of time. Even strolling among crowds of unfamiliar tourists generates a positive feeling related to curiosity and warmth. Similarly, some tourists sit at a window close to a river, watching leaves falling or other tourists strolling, so that they can feel some degree of self-contentment. All these feelings result from temporal autonomy, as these tourists can choose to stroll or sit, move fast or slow, and no one forces them to change the pace. For instance, Yehua (a female tourist from Beijing) describes how she spends time in Lijiang: "I naturally wake up at 11:00 a.m. and then play with the dog in the guesthouse. Around noon I stroll along streets to find a restaurant for lunch. After lunch, I stay in a café alone to read books for a whole afternoon." Yehua's sense of loafing highlights the way in which tourists stay away from clock time and follow their natural rhythms to organize their experiences. This sense in tourist destinations can be similarly found in the trip made by journalist Laurence Marks (2002) in the Maldives:

> I also learned another lesson by which I abide to this day: you don't need a watch. When I arrive I put mine in my suitcase and get up when it gets light, go to bed when it gets dark. I have no need to know what time it is. Who needs a watch in Paradise?

While Edensor (2012, p. 62) criticizes Marks for reinforcing a sort of romantic imagination that constructs postcolonial destinations as "a venue for the antithesis of the fast-moving, pressurized pace, and speedy lifestyle of Western, 'sophisticated' urban living," Marks presents a common tactic for many travelers, including my respondents in Lijiang, to experience alternative temporalities. As Adam (1998) asserts, machine time has been reified in capitalist societies to a point where people have lost touch with alternative rhythms and with the meaningful times of being in the world. Maybe in such a "paradise," either the Maldives, Lijiang, or other destinations, tourists can cultivate a sentiment of loafing, as long as they strive for alternative temporalities in a different spatial context.

The stories from Yue, Zhou, and Yehua point to positive feelings of comfort and relaxation among tourists in Lijiang Old Town. One common thread in these stories is that my respondents are female and travel alone. Among dozens of loafing-driven tourists I interviewed after 2015, female respondents express a stronger predilection for the town than their male counterparts. This gender difference deserves further analysis. In China, females work as hard as their male counterparts and furthermore, take the main responsibility for domestic duties. The time pressure on them is particularly stringent, which partially explains why so many female tourists travel to Lijiang for an alternative temporality. In a minor temper, Zhou emphasizes that "I don't care about how other people think of me, for instance, why I come to Lijiang so frequently or whether I feel bored after staying so long. Life is short and I need to enjoy myself in time. I want to do what I love to do." This quote alludes to gossip related to Zhou's frequent trips to Lijiang.

In reality, Lijiang Old Town has been known for one-night romances as a selling point to tempt male tourists. Many bars hire sexually attractive females to hook up with male tourists for consumption of expensive alcohol under the guise of a romantic encounter, all of which jeopardizes the town's image as a best tourist destination. For instance, in 2015, male tourists reported bumping into female hustlers, luring them to spend money in high-priced bars. A group of journalists from thepaper.com, an influential media platform in China, received these reports and went to Lijiang Old Town to investigate in October 2015. They noticed that bar hustlers were quite popular in the town and that many bars hired hustlers to tempt male tourists. One journalist noted that a female hustler indicated offering one-night sex service for CNY 1000 (~US$ 150).[1] Once released, the news item caused a large stir in China. The China National Tourism Administration stepped in to send a severe warning to the Lijiang Municipal Government. In a swift response, on 10 October 2015, Lijiang Heritage Authority took actions to suppress female hustlers. However, the link between one-night romances, female hustlers, and business fraud had by then been established to project the stereotype upon female tourists such as Zhou who frequent Lijiang for a different lifestyle. Although female tourists can feel free to travel alone in China, gossip about Zhou's frequent trips to Lijiang inevitably bothers her. Coming out of a patriarchal society in China, Zhou and other female tourists strive in their own ways to pursue their favorite lifestyle.

Another typical form of solitude that our respondents favor is to remain tranquil in a café to engage in self-reflection or watch passers-by. One sunny afternoon, I entered Prague Café, a place known for its loafing aura, and noticed that the background music was light and soft, and customers chatted in a low tone. Then I saw that Lihong (female, from Shenzhen) sat alone near a window. By having a cup of tea, Lihong took the opportunity to look out the window at tourists jostling along streets, or to gaze at the blue sky. I approached Lihong for an interview. As she explains:

> Lijiang is very suitable for loafing. To me, it means that in a café or a small pub, I steep good tea in boiled water and look out the window at tourists passing by. Meanwhile, I don't think about anything, but rather empty my mind. Hence I can fully indulge in the condition of real relaxation. Loafing is the best experience that Lijiang gives us. Look at the blue sky and white clouds! My heart is drifting with fluid clouds.

By looking at tourists and clouds, Lihong retreats into an unrushed moment of loafing in contrast to her hasty schedule in Shenzhen where she regularly resides. When I meet Yushi (female, from Hangzhou) in the Prague Café on another day, she had already spent a whole afternoon drinking tea alone. She talks about how she enjoys solitude: "In Lijiang no one knows me. I can completely relax among a group of unfamiliar tourists. In other words, I can be myself." Similarly, Yun (female, from Beijing) engages in contemplation to sort out her life journey: "Loafing enables me to empty or unload myself so that I can find a true self or discover a more positive self. There is no need to implicate myself in my terrible past. If I keep thinking about the past, I can feel very bad. It is time to look forward."

By loafing in the Prague Café or similar venues in the old town, tourists can calm down and remain tranquil, apart from the boisterous tide of tourist flows. Being tranquil has been regarded in traditional Chinese philosophy as an essential constituent of spirit. Whether spirit can be materialized is debatable, but the primary concern is to lay down the burden in order to obtain calmness for a while. Regarding self-reflection in the inner mind, pivotal figure in ancient Classical Philosophical Daoist Zhuangzi emphasized that "'Men do not look into running water as a mirror, but into still water—it is only the still water that can arrest them all, and keep them (in the contemplation of their real selves)."[2] Between running water and still water lies a mirror to reflect psychological activities. Here in Lijiang, these retreat tourists watch the flows of tourists, clouds, or water to contemplate on their own life and reach into their own mind, that is, into a home of their own.

The Prague Café offers notebooks to customers for personal reflection. In one notebook, I found some interesting notes written in Mandarin by domestic tourists: "In this café, I feel peaceful and calm, though other customers shout at each other and make noise. I really enjoy loafing in the midst of Lijiang's usual bustle of tourist traffic"; "I usually sit on a bench in Sifang Square [the town's center] and quietly look at other tourists. This is a moment of loafing, and also a moment of happiness." These notes offer a clue to what tourists prefer to experience in Lijiang. Various activities are conducted in order to stimulate a subjective experience of loafing – a time-space that enables tourists to temporarily keep a distance from time discipline.

Although my focus so far has been centered around female tourists, stories from male tourists equally deserve attention. When I met Pang (male, from Chongqing) in a guesthouse, he was applying a facial mask and holding a cigarette, with his back against a soft bench in the yard. In Chongqing, Pang is a skilled masseur, and his daily schedule is overloaded and busy: getting up at 10:30 a.m., starting his work at 11:00 a.m., and going to bed around 2:00 or 3:00 a.m. the next day. In Lijiang, however, he gets up and goes to bed at will, as he need not worry about work tomorrow. As he cheerfully mentions, he can turn off all alarms in Lijiang and wake up naturally:

> In Chongqing, I rent an apartment with a large bay window. My plan is always to sit on the bay to smoke and relax late at night. Then I realize that what I see through the window are high-rise buildings. No stars, no moon. My mood is gone and I feel very bad. Lijiang is a quiet place. Really quiet! What I can hear are sounds of the wind and flowing streams. This atmosphere enables me to easily relax my body and soul. Loafing is a wonderful moment of relaxation.

An overloaded work schedule in Chongqing exhausts Pang, while high-rise buildings alienate him from the natural world. In Lijiang, he retrieves his sense of nature, including wind and water, so that he feels that his life is not so uprooted.

All of these accounts illustrate a kind of temporal autonomy as respondents come to enjoy a retreat into a time-space where they extract themselves from everyday busyness, not through a heightened sense of time but through a nonproductive use of time. The stories shared by my respondents illustrate that an experience of loafing is associated with solitude through banal activities such as drinking tea and reading books so that they can manage time in a different atmosphere. These banal activities turn guesthouses and cafés in Lijiang into spaces of enjoyment, akin to the beach, an ideal site of leisure highlighted by Lefebvre. As Lefebvre (1991b, p. 384, original emphasis) reminds us, the beach generates a source of enjoyment for human beings because the body behaves "as a *total* body, breaking out of the temporal and spatial shell developed in response to labor, to the division of labor, to the localizing of work and the specialization of places." While relaxation at the beach illustrates the bodily source of authentic self, various places of solitude in Lijiang draw more on spontaneous self-reflection and banal activities. Nevertheless, both embody an escape from and a resistance to the cult of speed that underpins the process of capital accumulation.

Having said this, we must be careful about an absolute demarcation of temporal difference between Lijiang and elsewhere. There is no doubt that loafing is part of the hermeneutic circles of time, through which individuals swing between temporal autonomy in Lijiang and time discipline in their regular settings. Underpinning this swing is the logic of capital accumulation, a logic casting an enormous shadow upon tourists in Lijiang. Unsurprisingly, some tourists yearn for loafing, no matter how they understand what it is, because their mind is imbued with manic agitation and mental stress. Yali (male, Internet business owner from Beijing) frankly admits:

> I cannot become relaxed, no matter whether I am in Beijing or in Lijiang, since I cannot help but think about business affairs. Life is so stressful. I was diagnosed with a minor depressive disorder. It is not easy to survive in major metropolitan areas, especially for those migrants. Damn it, the high housing prices in Beijing! ... Beijing doesn't believe in tears, but in money.

Yali's complaint is not uncommon in Lijiang, showing an outcry against the race for wealth and prestige. Even in Lijiang, Yali cannot escape the manacle of time discipline, since he must cope with mounting pressure related to his business. Lijiang might present a different pattern of lifestyle, but to enjoy it or not depends on how much tourists can engage in a temporary moment of loafing.

Loafing through Socialization

While solitude is a favorite choice for some respondents, others engage in interpersonal activities to socialize with others. One popular activity is to drink tea together for casual conversation. Loafing is maintained through and experienced by a spatiotemporal pattern that seeks to minimize the ordering of time discipline and cultivates a cordial atmosphere, particularly in the space of consumption constituted by guesthouses and cafés, which set a leisurely pace and thus prolong the length of customers' stay. These settings are mobile, rhythmic elements in spaces of tourism consumption that become "meeting points" at which tourists congregate and interact for nonproductive use of time (Edensor 2012).

In Lijiang, every guesthouse dedicates one or two spots for tourists to get together for tea and conversation (Figure 4.1). The role of tea and chat in providing new experiences of loafing is closely tied to the creation of a sociable atmosphere among tourists who want to interact with each other, but stay away from serious obligation. Indeed, the tea table addresses itself to tourists as a friendly invitation open to tourists and even those passers-by, free from time discipline and explicit utilitarianism, from duty and obligation, so that participants around the table can enjoy a controllable adventure by drinking tea, chatting, and sometimes, simply listening. Sometimes business owners themselves provide tea service to tourists (Chapter 3), and sometimes tourists enjoy tea without the owners. Han (male, from Xiamen), who had already spent over two months in Lijiang at the time of our interview, says:

Figure 4.1 Tea table in Mahua's guesthouse in Lijiang Old Town in 2020. **Source:** Used with permission from Huang Qingyan.

> I haven't visited any attractions nearby. There is no need to visit them, partly because I don't have much money and partly because they are quite boring. Usually I drink tea and chat with other people in the guesthouse where I stay. In Lijiang, every guesthouse has a tearoom where tourists come together. Hence I can meet various persons and make friends with some of them. This never happens at home. In Lijiang, the most worthless item is time. People wake up naturally and get up in late morning. After brunch, it is time for me to drink tea. If no one is around, I read books and gaze upon the blue sky. Naturally I become idle.

Han finds an enjoyable experience in meeting other tourists for tea and conversation so that he need not be bothered by his forthcoming responsibilities. Indeed, tea and conversation are regarded as popular activities in Lijiang as a way to spend time. Liuyu, a male tourist from Guangzhou on his second visit in Lijiang, talks about his activities: "Loafing means doing nothing but staying in the guesthouse to drink tea or have casual conversation. There is no need to think about anything but just fun." More often than not, several tourists will saunter into a tearoom and gather together for tea, share their new findings about food and tea in the town, joke with each other, and laugh at the enormous pressure of living in big cities. The activity of chatting among tourists becomes a necessary strategy of socialization, albeit with no serious intent. Or if there is intent, it is about fun and relaxation, or "for the sake of communicating sociably with others without ulterior motive or serious purpose" (Foley 2017, p. 14).

Social activities such as drinking tea and chatting can happen in the big cities from which these tourists travel. Are these activities performed in Lijiang to express continuity with what tourists do in their regular abodes? Here, Nan (male, from Shenzhen) and Huahui (male, from Beijing) discuss how activities generate different meanings in Lijiang compared with their regular place.

Nan: In Lijiang, the main activities include chatting, drinking tea, and loafing .... In our own house, we live in a familiar atmosphere and are certain about everything. Here we do something we cannot easily do there, and thus get a different feeling. The reason that Lijiang is suitable for loafing is its slow pace and comfortable climate.

Huahui: Yes, I can drink tea, talk to people, and visit cafés in Beijing. But it is different. The longer I stay in Lijiang, the more I love it. Here I can forget the hectic life in Beijing and stay away from those annoying things. I have the chance to enjoy a simple way of life. That is why I have a different feeling here, even though these activities can be done in both places.

For both Nan and Huahui, talking to other tourists around the same tea table seems spontaneous, not for business, but for fun and enjoyment, though they rarely do so at home. One reason is that an experience of loafing is associated with a different rhythm and atmosphere so that these respondents do not feel obligated and pressured. While the rhythm is slow and calm, the atmosphere for tea drinking includes soft lights, paintings, comfortable couches, wooden tables, and entertainment magazines. Together, the rhythm and atmosphere make it meaningful to drink tea and chat with unfamiliar tourists, only in Lijiang.

Furthermore, gathering for tea and casual conversation stimulates a moment of "sociality." To Simmel (1997, p. 121), sociality denotes "the play form" of society, "the pure form, the free-playing interacting interdependence of individuals." This social interaction is bereft of content or substance since it possesses "no ulterior end, no content, and no result outside itself ... [and] is oriented completely around personalities" (Simmel 1997, p. 121). In other words, the interaction holds no utilitarian goals, but fosters a social game through which people play. The significance of play in everyday life is also highlighted by Lefebvre (2002) who emphasizes that the purpose is about relaxation and entertainment, and more importantly, a form of rediscovered spontaneity. Lefebvre (2002, p. 203) goes further than Simmel to argue that play is an activity "which is not subjected to the division of labour and the social hierarchies," though it remains marginal and sidelined in capitalist societies. As individuals are obsessed with "productivity, the cult of speed, and a fear of wasting time" (Ehn and Löfgren 2010, p. 11), these ordinary activities such as tea drinking and casual conversation in Lijiang have no bearing on productivity but are engaged in simply to spend time, and thus seem, ironically, *extra*-ordinary.

For these tourists, social interactions around tea tables tend not to become utilitarian and the imperatives of production and efficiency are temporarily suspended to orchestrate the use of time and space. In big cities, the optimization and acceleration of time use holds sway over the nonpurposeful use of time, but in Lijiang, the dominant logic of time discipline is reversed. This choice concurs with what Lefebvre (1991b, p. 87) calls a project of radical disalienation, or "a rehabilitation of everyday life, shedding new light on its positive content." While Lefebvre labels this project as *autogestion*, or self-management, loafing entails similar content regarding spatial and temporal autonomy. Having said this, loafing in Lijiang is geared towards a temporary palliate rather than a permanent rehabilitation, concerned with the effective escape from social acceleration and spatial alienation. This palliate can be made possible through commercial transactions and social division of labor in a well-organized tourism industry. It is part of modern capitalism from which retreat tourists want to escape temporarily. In other words, loafing becomes a means to adjust the time-space of life to participate better in the race for wealth and fame.

Furthermore, socializing for loafing is fleeting. Strictly speaking, tourists and business operators maintain a degree of respectful distance as they can unfetteredly talk about here and now, but try not to unravel too much there and then. Tourists socialize in the guesthouse in anonymity, and maintain a form of unacquainted friendship with each other, since they have no conflict of interest and will remain detached once they leave the guesthouse for their regular cities. Before coming to Lijiang, tourists do not usually know each other; around the tea table, they feel comfortable in the company of unacquainted persons; after the trip, their friendship fostered in Lijiang generally vaporize and they yearn for the next adventure, to meet new friends for casual company.

The analysis of loafing suggests a break from the heightened sense of time brought by tourism, and thus calls for the need to de-exoticizes tourism. As Franklin and Crang (2001, p. 8) argue, tourism studies have privileged "the exotic and strange, reflecting anthropological legacies, to speak of dramatic contrasts between visitors and locals." Loafing through banal activities depicts a sort of temporal autonomy, where the usual busyness in everyday life can be suspended and the norm of "life as a race" no longer applies. Thanks to loafing, tourists reside in a condition full of possibilities and ordinariness, doing things that they usually have no time or mood to do in their regular environs. In other words, they can search for a different moment of being-in-the-world as a subjective reaction to "the condition of efficiency and speed" imposed for the exhaustive use of time for productivity (Foucault 1979, p. 152). They follow natural inclinations in their experience of the passage of time, without succumbing to the clock, so that the social meaning of time becomes more ordinary than ordinary life. These touristic experiences of loafing involve a self-conscious choice of consumption to resist time discipline, albeit a choice that is facilitated through and conditioned by tourists' pecuniary ability. Through these experiences, tourists can keep a distance from social constraints (time discipline, obligations, etc.) and suspend clock-driven temporal routine so that they can experience cyclical time.

The contrast between Lijiang and China's big cities in terms of tempo is stark. In big cities such as Beijing and Shanghai, my respondents report that they turn every minute into an endless race to the finish line of material prosperity. They are unable to grasp the world as a meaningful entity due to ubiquitous sociospatial alienation. In Lijiang, they experience a quite different temporality as they turn every hour into a play of doing nothing, or doing something without purpose. Neither a finish line nor a rush moment comes to bother them. Banal activities such as tea drinking, chatting, gathering for a meal, and so on, can promote interpersonal relations and contribute to a sense of home. As a result, Lijiang represents a local resistance against the global cult of speed. By slowing down and enjoying loafing in Lijiang, some tourists find that they can free themselves from this cult and rejuvenate from physical exhaustion and mental stress. Only in Lijiang do they unplug the clock for a retreat.

Regarding Lijiang Old Town as a Home

To tourists, a sense of home in Lijiang entails two layers: guesthouse as a living space and the town as a loafing space. Given that homes "always involve encounters between those who stay, those who arrive and those who leave" (Ahmed 1999, p. 340), the guesthouse in the old town exemplifies a fluid dimension of home that entails increasingly normative journeys of stay and departure. Bedrooms in the guesthouse are not a main focus when tourists talk about how they feel about Lijiang. The necessary supplies, including soft warm bed, refreshing water, cleanness, and cozy decoration, helps tourists to escape from anxiety and even insomnia, and thus pursue sound sleep for mental and bodily relaxation. Bedrooms do reflect the business operator's taste of and efforts toward homemaking, but have no bearing on tourists' personality. One key function of the bedroom is to enable customers to enjoy sound sleep and wake up naturally.

Unsurprisingly, the tourists I interview do not treat bedrooms as a focal point of their stories in Lijiang, although they do appreciate that their sleep is much better in Lijiang than in their regular cities. They talk more about their personal reflections and sociable activities in guesthouses. Different from a hotel where its bedrooms remain the center of activities, a guesthouse in Lijiang spotlights its semipublic spaces more than its bedrooms in order to foster socializing between guest and host and to facilitate access to the natural environment. Thus, the semipublic space in the guesthouse is for sociable encounter, without worrying too much how time is spent or why. Due to this sort of encounter, tourists are welcome to explore new possibilities of things, people, and information, and thus they are pulled out of their routine activities.

As an abode for dwelling and socializing, the courtyard guesthouse in the old town shows itself not merely as an environment for sound sleep, but also as the comfortable aura where individuals can rest their soul with loafing and comfort. To create a homely setting for tourists, guesthouse operators aspire to foster a home away from home, not through standardized procedures implemented in starred hotels, but through personalized service and facilities (see Chapter 3). Homely life is expressed in the guesthouse through music, tea, conversations, meals, or simply enjoying sunshine and flowers in the courtyard. Furthermore, this life is bereft of domestic chores and family obligations (see Chapter 5). These elements reify the representation of ideal home for tourists – beautiful, natural, and comfortable – so that tourists are willing to spend money in the guesthouse and relax themselves. In terms of service, high-end guesthouses allow customers to have free access to fruits, kitchen, tea, and snacks. This kind of freedom enables tourists to really feel at home. Here I cite two quotes from tourists when they think about home in Lijiang old town:

> Once I visit Lijiang, I choose this guesthouse. It gives me a sense of home. It provides necessary services to make me feel comfortable. The boss is also very nice, and creates a very relaxing and pleasant living condition. It has a deck looking out to the old town, supplies free fruits, and provides books. I can spend a whole day basking or loafing in the guesthouse. (Zhou: female, from Wuhan)

> I enjoy a life of ease in Lijiang. When I am hungry, I go to the guesthouse's kitchen to find food; when I am tired, I return to my bedroom to fall asleep. My whole disposition is one of contentment and ease. There is no real restriction. When I stay in Lijiang, I pretty much forget time. I don't care about how time is passed in Lijiang. (Hai, female, from Guangzhou)

For a tourist, what does it mean to be at home in a destination? To Zhou and Hai, the familiar place in Lijiang, the place that they frequent and stay with comfort, embodies a feeling of being at home. Such a place embodies a sensed home, partly because they are familiar with it through multiple visits and partly because it fulfills their imagination of an ideal home. They know that they must return sooner or later to another place where they have regular abodes, but here at this moment they can relax among a group of unfamiliar tourists and escape from time discipline. Lijiang literally becomes the some-where and elsewhen of home, which retreat tourists temporarily inhabit for a transient moment of belonging, a moment that expresses the very logic of existential experience, that is, their very subjective feeling about cordial interactions with the town's social and natural environment. Their sense of home is unrelated to memory, nostalgia, or shared past, but draws on their own imagination and experience. That is why they and many other respondents keep reiterating "I feel" or "I have a feeling of."

The unique combination of comfortable aura and homely service acts as a catalyst to foster a sense of home among tourists. Customers do not feel a sense of strangeness in the guesthouse, as everyone seems to be the host, not through a perspective of property ownership, but through a sense of spatial and temporal autonomy. Even the business operator merely leases the courtyard house from native residents and temporarily runs it for profit and lifestyle as well. Here, the guesthouse helps tourists to live with spatial and temporal autonomy because it exemplifies what it means to be at home, but free of the domestic burdens of their regular house and of a hectic work pace that consumes their every minute. The relation to the guesthouse expands beyond sound sleep and waking up naturally, as the guesthouse creates a homely setting that enables tourists to indulge into either lonely retreat or light-hearted socializing. The anchoring of the guesthouse occurs both spatially and experientially through its homely aura, that is, a home away from home, to offer comfort and retreat to those exhausted in the relentless race for wealth and fame. Thus, the aura is associated with feelings of relaxation, security, and being settled. Through loafing in the guesthouse, tourists seem to bring their own time and space under control, as they are not bothered by any order from someone else.

A sense of home can expand from the guesthouse to the entire old town. Tourists choose to become wanderers in a town where no one knows them. Strolling in various alleys and streets, they can get lost and then figure out their way to Sifang Square, the town's center, to get their bearings. A feeling of ease permeates their walk so that they can spend time and explore space free of utilitarian purposes. It is this feeling that frees tourists from the centripetal force of time discipline in their normal life and enables them to experience an alternative time-space with other people through friendly, yet ephemeral, ties. During their trip in Lijiang Old Town, tourists retrieve the social meanings of their time and rebuild connections with their surroundings beyond the walls of their regular life. In other words, the trip reveals something of a homely lifestyle, of a way of life in the original sense of self-fulfillment, a sense which unfortunately has been lost in a race for wealth. This sense of fulfillment highlights the importance of Lijiang Old Town as a place where tourists can retreat from the hectic pace of modern urban life.

Different from the "home" they leave behind, the sensed home that tourists experience in Lijiang does not require commitment or responsibility. If they do not enjoy socializing with other tourists and guesthouse operators, they can stroll alone in the town; if they prefer to hang out with others, they can participate in relaxed conversations about food, tea, and Lijiang's interesting attractions. Without commitment or responsibility, they develop a free-floating experience in the town. For instance, the comment made by Hai is quite typical among tourists: "In other words, I hope to enjoy a sense of home, but don't want home-like pressure. My own home is about daily life with various pressures. I come to Lijiang for a different living condition. It is about ease and comfort." This utterance generates two propositions related to homemaking. First, if home represents some degree of attachment and bond, then tourists in Lijiang want to pay for a sense of home without making any commitment in time and space. This sense alludes to experience economy, by experiencing home first and fast in Lijiang, in order to address their loss of home feelings in big cities. Second, the burden of homemaking is shifted to female service workers who come from villages around Lijiang. Such a division of labor can liberate tourists, particularly female tourists, from various constraints that society has imposed upon them, including domestic chores, key performance indicators, punctuality, and overwork. As I will discuss in detail in Chapter 6, tourism development in Lijiang implicates female workers from rural Lijiang into highly differentiated and unequal positions in the tourism industry in general, and in providing homely service to tourists in particular. The synthesis of class difference and gender disparity reinforces the gendered division of labor in the process of homemaking by enabling retreat tourists to enjoy a homely milieu in the guesthouse, while shielding them from domestic burden.

Retreat tourists follow their heart to experience time and space, becoming pretty much traditional Chinese loafers such as Tao Yuanming, in order to get closer to nature – both internal and external. Nevertheless, wild nature, remote in mountains and rivers, seems overwhelming for comfortable retreat; the humanized nature, embodied in Lijiang Old Town's waters and flowers, provides a safe and enjoyable setting for retreat tourists to temporarily enjoy loafing. Similar to Tao Yuanming who reclused himself in a piece of farmland to enjoy trees, dogs barking, the cock crowing, cooking smoke, and the southern mountain (Tian 2005), these tourists relax in the old town's neat courtyards to enjoy flowers, floating clouds, tea drinking, and so on. The difference is that Tao was surrounded by a natural environment transformed for agricultural plantation and rural life that Fei Xiaotong (1992) described, but retreat tourists in Lijiang dwell in a humanized environment that is sustained by constant care (see Chapter 6) and modern facilities. My respondents seem not to question the meaning of culture or challenge the authenticity of Naxi heritage landscapes, but rather treat the town as an oasis to which they can escape from time discipline and social alienation. Enjoying spatial and temporal autonomy elsewhere is definitely secular, and yet deeply spiritual. While retreat tourists have no opportunity or ability to follow Tao Yuanming and other reclusive scholars in ancient China to give up the pursuit of wealth and fame, they use their purchasing power to stay in Lijiang for a temporary retreat. This retreat is not about the resolute reclusion that Tao Yuanming embraced, but rather a distance from the cage of time discipline so that they can weave themselves into the cyclic way of living.

Indeed, retreat has become a luxury enjoyment at very high cost of time and money, and not everyone can afford it. The joy of feeling at home in Lijiang, strictly speaking, is reserved only for those who have adequate purchasing power in a consumerist society, but yearn for a retreat from the hustle and bustle of urban life. In China's big cities, these property owners give up a cultural home to maintain a commercial house. Once they feel culturally homeless, they choose to visit Lijiang and other similar destinations for a short retreat into a different time-space. The making of sensed home is particularly explicit in the terrain of tourism, as tourists make efforts to embed themselves in destinations for a transient experience of familiarity and attachment. Tourism enables "different modes of attachment to and detachment from places, cultures and people; different modes of belonging in the world; hence we may conceive of tourism as a way of positioning oneself in the world" (Haldrup 2009, p. 54). Here and now in Lijiang, retreat tourists enjoy the Lijiang lifestyle and claim a sense of belonging to the town so that they can engage in a break with the feverish monotony of the everyday (Lefebvre 2002). Afterwards, they resume their hectic pace of living in big cities and Lijiang might become a memory to provoke their homesickness. Maybe their actual abode is not the best approximation of their ideal home, but a permanent departure for a culturally constructed dream home elsewhere is not necessarily in their best interest. As a result, they enter into an unhomely condition, a topic I discuss next.

Being Unhomely in a Mobile World

As I noted earlier, the slow and qualitative forms of relaxation become a temporal strategy to leave behind the hectic mundane life at home. The spatial difference between Lijiang and big cites is temporally reflected through an explicit distinction between loafing and busyness, relaxation and stress, and being unloaded and overloaded. By staying in Lijiang for a while, these tourists hope to both escape from hustling cities and recharge their body and mind so that they can better handle urban life once the respite is over.

To these tourists, loafing in Lijiang represents an art of living with spatiotemporal autonomy. They are willing to keep distance from the hectic pace of urban life, albeit in a very temporary manner, and detach from value-driven production. The experience of loafing implies a rejection of what Rosa et al. (2017) call "the escalatory logics of capitalist modernity." Time is not wasted for doing nothing, but enjoyed as active, yet tranquil, performance. This intentional immobilization of time makes them different from other tourists who rush to visit various attractions, go shopping, and take photos. Here, to experience loafing means to come and stay without busyness, so that they can enjoy their time with active retreat and slow savoring. It leads to the authentic ecstasy of unproductive time-space. Nevertheless, we cannot enshrine loafing, as these tourists need to return to their busy pace after an ephemeral stay in Lijiang.

For my respondents, tourism for the sake of loafing is made possible by their pecuniary ability. The ability to travel to Lijiang and stay in a guesthouse actually comes from tourists' involvement in the process of capital accumulation and compliance with the cult of speed in their regular abodes. They realize that they enjoy a temporary stay in Lijiang, but understand the need to return to their busy daily life. The routinized pace, which consolidates time discipline, intersects with experiences of loafing in Lijiang to reflect what Lefebvre (2004) calls a dialectical relation characterized by unity in opposition, giving rise to an unhomely life journey between stay and departure. In exploring this journey, assertions abound concerning the spatial representation of Lijiang. When asked whether she wants to quit her job and migrate to Lijiang for constant loafing, Lifei, a female tourist from Chengdu, rejects the possibility by saying: "After all, traveling requires necessary economic capital. I even think about finding a job in Lijiang. Nevertheless, the salary in Lijiang is pretty low. Another way is to run a shop, but I heard that the tourist business is not rosy. Thus I need to go back to Chengdu." Similarly, Zhou rejects the possibility of migration, but leans toward repeat visits, as she explains:

> I did seriously think about migration to Lijiang. After considering many factors, including food, public services, and transportation, it is not my choice. If I settle down here year-round, my love for Lijiang might be changed. If I turn Lijiang into a site of career and earning, the original feeling that it gives to me would be changed, too. Nevertheless, it can become a second-home site.

Even though Zhou acknowledges that Lijiang deserves her full affection, it is a destination where she can enjoy loafing as a tourist, rather than a place to settle down as a resident. For both Lifei and Zhou, social and economic factors in Lijiang discourage them from a loafing lifestyle, but their pecuniary ability acquired through busy work in the big cities allows them to enjoy a taste of temporal autonomy in Lijiang.

The struggle between stay and departure described by Lifei and Zhou highlights tourism consumption not as an escape from and a resistance to mechanisms of capitalist accumulation in time and space, but as a form of negotiation between the self and the world in order to carve out a spatial and temporal pattern that enables tourists to adjust to the established order. The struggle is captured well by MacCannell (1973, p. 589) who describes the concerns of moderns for "the shallowness of their lives and inauthenticity of their [daily] experiences." To overcome this shallowness, individuals become tourists to engage in the pursuit of authenticity elsewhere, although this pursuit is always based on wishful thinking and even romantic imagination. While tourists are supposed to experience authenticity in destinations, my analysis alludes to a different dynamic related to busyness in daily life and loafing in destinations.

In addition, temporal autonomy brought by loafing exists under the shadow of a cruel reality. Although tourists do not care much about personal background when they gather together for tea and conversation, social status differences do exist. Nan (male, thirties, from Shenzhen) categorizes tourists in Lijiang into two types: "Lijiang is a place for winners to enjoy a relaxed lifestyle with a slow pace, and for losers to escape the harsh reality. Both types of tourists can temporarily leave daily things behind." This categorization is widely recognized among tourists and tourism operators in Lijiang, indicating ambivalent meanings of loafing. To Duo, a male tourist from Shanghai, loafing is more like an illusion than a reality, simply because of the materialistic pressure he feels:

> Enjoying Lijiang lifestyle depends on whether you have resources. Today people face mounting pressure and it is quite rare for them to become idle. Loafing requires a good state of mind. Even though there is no economic pressure, mental stress does exist. How can tourists really enjoy this beautiful landscape in Lijiang? It does cost much money, doesn't it? If only I could truly enjoy it, my mind would stay calm and peaceful. This is what I call loafing. I don't think I can thoroughly relax myself. I am just too exhausted to enjoy relaxation.

Even though he can relax in Lijiang, Duo cannot fully leave daily affairs behind. His utterance "too exhausted to enjoy relaxation" indicates that temporary loafing in Lijiang cannot calm him down, and there is no single panacea for the ills of speedy modernity.

Physical distance from their regular abode cannot cut off tourists' ties with job-related interruption. They still receive phone calls related to work, no matter how they use a holiday in Lijiang as an excuse for temporary detachment from business and duty. The continuity of busyness is most starkly expressed by Yali: "In Lijiang, I don't have much pressure, but I do have to handle annoying things. Work-related affairs reach me by phone. It is annoying." Since he must handle urgent business affairs in Beijing, the pursuit of loafing in Lijiang seems similarly illusionary. One call after another in Lijiang deprives Yali of the mood of relaxation to such an extent that he keeps busy, only in a different geographic context. Caught between loafing and busyness, Yali and other respondents cannot engage in a full meditation by forgetting business affairs and economic pressure in their regular cities, but have to endure an unhomely condition.

The final aspect of ambivalence is the dynamic relation between slowness and busyness. Tourists take the chance to engage in loafing and thus recuperate from physical and mental stress. Lijiang becomes a space of rejuvenation, or a site for retreat from which these respondents can launch a new life journey. My interview with Liuyu and Heng, a couple from Guangzhou, demonstrates that loafing in Lijiang and busyness in big cities could be two sides of one coin in the context of China's speedy modernization. As they put it,

> Life should strike a balance between busyness and relaxation. After spending some time being relaxed, we can have better work efficiency when we return. So the purpose is to release some of the negative effect. (Liuyu)

> To be frank, we don't want to return to Guangzhou. We feel so comfortable and relaxed in Lijiang. Nevertheless, we can resume our normal life after returning to Guangzhou. We will definitely revisit Lijiang, since we have such a fabulous impression. (Heng)

By engaging in loafing, Liu and Heng demonstrate that they have tried an alternative temporality through which they feel good and recover from overloaded work. This engagement is very transient, as they need to return to the bustling city in order to earn money for loafing in Lijiang or elsewhere. Similarly, Luo, a male tourist from Beijing, notices the necessity to swing between loafing and busyness: "Everyone chooses to escape high stress in big cities and retreat in Lijiang. They want to stay away from their regular lifestyle since they cannot live forever with the stresses and strains of fast-paced life in big cities." This juxtaposition between Lijiang and regular settings, between loafing and busyness, reflects what May and Thrift (2001, p. 19) argue, that one cannot have "a speeding-up without also having a slowing down." The same holds true when tourists can better handle their bustling life in cities after they relax in Lijiang for a while. Loafing is for the sake of busyness; similarly, busyness loses its ground without loafing. Through loafing in Lijiang, tourists seek an acceptable (and comfortable) way forward: not a complete passage that allows them to discard busyness, but a purposeful search for retreat and autonomy.

Nevertheless, loafing as a tactic aims not to destabilize the dominant culture of speed, but enables tourists to live with it in a relentless world. Thanks to loafing in Lijiang, tourists can return with energy and refreshment to their home and resume hectic lifestyles in preparation for loafing during future trips. They remain complicit with a hectic lifestyle with short moments of loafing, which differs from a balanced way of life that Chinese ancient philosophers envisioned. Luo (male, from Beijing) feels that

> [i]n big cities, you cannot slow down. If you do, you find you are out of place. You feel a sense of identity crisis and even a sense of fear. When you come to Lijiang, you need to slow down because everyone else enjoys slowness.

Loafing in Lijiang becomes meaningful and desirable only in relation to busyness in big cities; otherwise, the nonproductive use of time degenerates into laziness and indolence. Therefore, tourists oscillate between here and there, busyness and loafing, a dynamic process adding new spatial and temporal meanings to home-making. The oscillation echoes Bergmann's (2008, pp. 14–15) argument that the use of time is fractured by two contradictory understandings: "unlimited freedom of speed and acceleration on the one hand, and the vision of slowing down and slowness on the other." While Bergmann emphasizes that these two understandings have not been reconciled with each other, tourism for loafing might aptly address his concern. Tourists engage in busyness so that they accumulate the pecuniary ability to travel for loafing, while their experiences of loafing in turn lessen their bodily fatigue and mental stress so that they can better participate in the hasty process of capital accumulation.

The relationship between loafing and busyness, slowness and speed, is not mutually antagonistic or exclusive. To tourists, loafing in Lijiang represents a manageable and even palpable way of living with speedy modernization. They cannot permanently escape the logic of time discipline, but they can modulate it to their ends through a nonproductive consumption of time elsewhere. Tourism allows them to step outside the everydayness of social acceleration, without actually leaving it. After a temporary stay in Lijiang, these tourists move back to daily hectic rhythms and live with homesickness toward their next trip. They work hard to earn a loafing moment in Lijiang (or elsewhere), and this moment has one key purpose: to get away from busyness and then get ready for busyness. This becomes a vicious circle, and very few can go beyond it. The ambivalence of loafing, I argue, can be approached as a barometer to reflect "a deep-rooted and ever-increasing malaise" that foregrounds the deal of modernity (Lefebvre 2002).

Conclusion

This chapter has explored one specific temporal element of tourism – loafing – and how tourists engage in a nonproductive use of time for retreat. In Lijiang, retreat tourists engage in intrapersonal and interpersonal loafing, and must handle the swing between loafing in Lijiang and busyness in their regular abode. Their moments of loafing are constituted by banal activities: solitude (waking up naturally, meditation, and watching passers-by) or socialization (drinking tea with others and chatting for fun). In addition, these tourists develop an ambivalent feeling toward loafing, as shown by their love for Lijiang, but refusal to migrate there after careful calculation, or by the fact that their pursuit of loafing is derailed by economic pressure. Retreat tourists must reset their temporality again and again, although the discourse of speed casts them as vulnerable, open to intervention. They are expected to fit into a dominant temporal order of speed and efficiency in big cities, but harness their purchasing power to escape elsewhere for an alternative temporal style of relaxation and slowness. As Sharma (2014, p. 18) asserts, individuals in a relentless world must know how to recalibrate different temporalities, or learn "how to deal with time, be on top of one's time, to learn when to be fast and when to be slow." To recalibrate time is not merely about synchronization of personal paces with capitalist temporal ordering, as Sharma (2014) emphasizes, but also the possibility of retrieving the meaning of time from the race for wealth and fame so that individuals can feel at home, no matter how ephemeral this feeling may be.

By participating in certain routine activities in the guesthouse as well as in the town, these tourists can make themselves at home in a destination where they spend days or weeks for retreat. Guesthouses and the town become their home away from home, embodying the living of time that Blunt et al. (2021) associate with the temporalities of urban dwelling at home, which entail both staying put and moving on. As Blunt et al. (2021, p. 160) note, the living of time in cities reflects "intertwined migration, housing, and family histories that are bound with, and articulated through, narratives of domestic and urban (in)stability." While Blunt et al. (2021) note that urban homemaking is deeply shaped by housing precarity, which creates structural contingencies of tenure, property ownership, and displacement, they also identify the value of home-city geographies to unravel the prevalence of unhomely life among those who cultivate a sense of rootedness in a turbulent housing market and who may be physically rooted yet feel a strong sense of affective displacement. And Blunt et al. (2021) use housing precarity as a primary context, I notice that retreat tourists in Lijiang often care more about the living of time and space for an experiential sense of home. Perhaps they own properties in big cities so that in Lijiang they feel unburdened by housing precarity. Different from regular urban residents who, according to Blunt et al. (2021), seek a sense of long-term belonging in response to the unpredictable present, retreat tourists come to Lijiang to experience short-term spatial and temporal autonomy, whereby they escape from and adapt to time discipline. In this regard, retreat tourists' transient sense of home stands for escape, rejuvenation, and comfort, as opposed to long-term residents' aspiration to a feeling at home, which can stand for rootedness, placement, and mooring. Furthermore, these tourists are in the privileged position of being able to make themselves at home in Lijiang, and can return to another "home" at the end of their journey. This privilege is heavily inflected with class-based purchasing power and a geographical imagination of alternative places and cultures, more or less drawing on Han-centric ideals of ethnic cultures in China's periphery.

Due to tourism, the desire for temporal autonomy comes true in a different spatial context, through which tourists can enjoy reflection and restoration. Loafing helps retreat tourists in Lijiang to relearn how to cherish and experience a relaxed sense of time. The quest of all modern thinking on time-space, as May and Thrift (2001, p. 37) specify, has been to "be filled by and to amplify the presence of the now, to make the present habitable and visible by remaking what counts as past and future, here and there." While a condemnation of speedy modernization cannot ease pent-up anxiety about busyness and acceleration, loafing in the time-space of tourism might generate new opportunities for both making the present habitable and reconsidering the role of tourism in wider debates on everyday life. To these tourists, loafing in Lijiang entails a temporary respite from busyness in China's big cities, which allows them to handle better the relentless race for wealth. To make it possible, a certain degree of spatial and temporal detachment from their regular setting is needed, and this detachment can come into reality only if these tourists process the pecuniary ability to afford transportation, food, and lodging. The hiatus facilitated by a temporary stay in Lijiang Old Town can generate a sense of home that serves as a complement to the hustle and bustle of urban life. A swing between loafing and busyness can make livable these tourists' hectic lives in China.

Travelling to Lijiang for a sense of home can, of course, be read as an expression of resistance to the evils of modernity in the context of China. It allows tourists to play with an alternative lifestyle in ways that may, for many middle-class people in China's big cities, change the common understanding about money, time, and the meanings of life. Yet, to propose that this retreat in Lijiang represents an overt or covert opposition to the Party state is to lose sight of far more nuanced and interesting implications about China's sociospatial transformation. I would argue instead that the consumption of a loafing lifestyle in Lijiang Old Town is part of a larger pattern of intentional deal made by individuals between money and meaning. Like those who choose alternative lifestyles by migrating to other countries, retreat tourists embrace the enrichment of daily life allowed by their purchasing power, learn to adjust their temporal patterns elsewhere, and swing between loafing and busyness. Unlike those transnational migrants who leave Chinese big cities for foreign countries, these domestic retreat tourists seem to find, rather than discard, a sense of home born out of consuming alternative lifestyles without abandoning their routines in a heartless world.

How to deal with time, or practice what Confucius calls the balanced ways of King Wen and King Wu, remains a key challenge to individuals in our speedy society (Angle 2009). Here we detect a dialectic of adjustment between loafing and busyness, between here and there, which serve as tactics, a term from de Certeau (1984, p. xxiv), that from the perspective of individuals, instead of institutions, entail everyday practices to cope with "the fleeting and massive reality of a social activity at play with the order that contains it." This dialectic adjustment permeates the everyday life of which tourism is a part. This observation sounds pessimistic, corresponding with a more dreadful illustration of the consumer society by the Frankfurt school scholars, but exemplifies how individuals live with this cruel and fluid reality. Indeed, the perception of loafing and busyness is not a matter of choice, as these two tempos mutually define each other. Lijiang Old Town happens to be a temporary stop for individuals to rejuvenate their sense of home so that they can endure an unhomely life journey. Hence, it is necessary not to treat home and homelessness as a duality, antagonistic to each other. Rather, they implicate one another in a personalized way and their unity unfolds an endless journey of arrival and departure. In Chapter 5, I shift to native residents in the old town and reveal how they handle the deal of modernity between money and meaning in the process of tourism development. Between departure and stay, the making and unmaking of rooted home unfolds in their struggles and adjustments.

Notes

1 https://www.thepaper.cn/newsDetail_forward_1385622

2 Original text: "人莫鉴于流水，而鉴于止水，唯止能止众止" 《内篇-德充符》 The Inner Chapters: *The Seal of Virtue Complete*, translated by James Legge. Source: https://ctext.org/zhuangzi/seal-of-virtue-complete/ens

Chapter Five
Displacing Native Residents: Money, Meaning, and the Remaking of Home

Displacement and uprooting are an integral part of human experience, regardless of class, gender, or age. While much has been written upon how marginalized people suffer from forced eviction, little attention is paid to those who have adequate resources to secure a decent abode, but must endure a loss of home in their everyday life. If we recognize a settled way of life as the kernel of a homely environment in the world, then the condition of displacement at home can generate an unsettling effect upon how individuals experience and think about the practices of everyday life.

Writing on the dispossession of urban space, Harvey (2010b, pp. 247–248) argues that "questions of security, fear of others, social preferences and prejudices all play their part in the fluid conflicts between social groups over the control of space and over access to valued assets." These questions are particularly true to those vulnerable in the process of either gentrification or forced eviction, or both. Their experiences of precariousness, uncertainty, and waiting are reflective of "widespread fear of dispossession," economically and culturally (Harvey 2010b, p. 248). The literature on gentrification documents practices of harassment, coercion, and intimidation imposed upon residents in gentrifying neighborhoods and how these practices make homes unhomely (Huq and Harwood 2019). Some tenants are displaced by illegal means and become victims of forced eviction. Nonmarket intimidation strategies, coupled with class and racial inequality, further reinforce the structure of domination and marginalization and exclude marginalized groups from the sites they have called home. Thus, Betancur (2002,

Unhomely Life: Modernity, Mobilities, and the Making of Home in China, First Edition. Xiaobo Su.

p. 806) argues that the home, as a sanctuary against all the instances of social and economic inequality, is not immune from "extreme forms of manipulation of the real estate market through racism, abuse of public office, and utilization of criminal and other 'nonmarket' intimidation strategies."

What happened in Lijiang Old Town is akin to Harvey's analysis, but in a different way. Due to tourism development since the late 1990s, market forces and rising property values have played an instrumental role in converting courtyard houses in the old town from residential uses to tourist consumption. Native residents' struggles that ensue in the process of conversion reflect an imminent crisis about who has the right to live in this increasingly crowded town and who can call it home. As native residents must cope with mounting negative impacts brought by tourism development, to stay in the town or move out becomes a fraught question. This question is not only about the rational calculation of property values and economic returns, but also more deeply about the changing meanings of home and the representation of Naxi ethnic culture.

Framing Lijiang Old Town as yet another case of tourism-driven gentrification is reasonable, but this generalist reading does not tell us much about how displacement and a loss of home are experienced and embodied by those who were actually born and grew up in a place so dear to their heart. Whether they stay in the town or leave for the new city, native residents in Lijiang do not cope with the fear of dispossession, but keep a tight grip on property ownership and make decisions on their own. Increasing numbers of residents choose to lease their property to migrant businesspersons in exchange for good economic returns. Those who choose to remain in the old town establish a proprietary sense of ownership over, and belonging to, the town even as they handle an uncanny condition of in-situ displacement.

Displacement does not happen exclusively to those leave due to forced eviction or housing demolition, but happens also to those who stay in situ, yet feel alienated from the place where they live (Kern 2016; Spangler 2020). As Gupta and Ferguson (1992, p. 10) point out, "for even people remaining in familiar and ancestral places find the nature of their relation to place ineluctably changed, and the illusion of a natural and essential connection between the place and culture broken." In New Orleans, for instance, Airbnb has penetrated many neighborhoods to promote a mixture of tourists and residents in ordinary living spaces. Residents feel put on display to unfamiliar tourists in order to demonstrate an authentic life through dog walking or casual chatting. Nevertheless, this display makes them feel uncomfortable due to a loss of control in their ordinary spaces of life when confronted with "the material effects of Disneyfication" (Spangler 2020, p. 581). These local residents produce the very homely milieu of urban neighborhoods that Airbnb attempts to promote among potential tourists, but become displaced in situ. In other words, these residents perform emotional labor for free and contribute to the process of value production shared by Airbnb, property owners, and tourists.

In Chinese cities, urban redevelopment is often filled with violent evictions and dramatic demographic changes. The prevalent symbol of the Chinese word "拆" (chai, literally, demolishment) in many old neighborhoods in these cities epitomizes the brutal dispossession of urban space (Lam 2022; Wang 2022). In Lijiang, however, in-situ displacement is very ordinary, noncatastrophic, and subtle. Indeed, this form of displacement can be read as symptomatic of the ways in which the tourism-driven logic of profit making redefines the function of courtyard houses, as well as the social life sustained within these houses. The function of courtyard houses undertakes a transformation from the use value of daily dwelling to the exchange value of profit making. Accordingly, well-established neighborhoods, and the ways of living in the old town, are in dramatic flux. Thinking through this transformation, which unsettles the representational and experiential dimensions of the home, calls into question the battle of making home for one group of people and unmaking for another. By analyzing their struggle between stay and departure, this chapter exemplifies experiences of displacement that native residents have undertaken, without simply assuming homemaking as a business of purchasing a property for dwelling and ownership.

An analysis of homemaking and unmaking by native residents in Lijiang Old Town articulates the social and cultural consequences brought by massive urban redevelopment in Chinese cities. This analysis does not lead to an acritical reading of China's great transformation that has uprooted many individuals, but rather unravels how the deal of modernity is embodied within the practices of everyday life by those who have certain agency to make a choice. Thus, this chapter generates a nuanced understanding of home, and of the struggle for rooted home in a relentless world, calling attention to a dramatic transformation from familiarity to strangeness caused not by forced eviction, but by voluntary move-out under changed circumstances and displacement at home. The transformation is not derived from the typical process of capital accumulation by dispossession per se, as we find in many cases of gentrification, but explicitly reflects how the deal between money and meaning shapes the process of homemaking.

A Sense of Home in the Town

Many elderly native residents, including those I interviewed between 2002 and 2010, felt reluctant to move out and give up their ancestral courtyard houses in the old town. During this period, I noticed that some young and middle-aged adults did stay in the town, but for tourism business rather than pure residence. As long as they could afford to buy a house or an apartment in the new city, younger native residents moved out for a modern lifestyle. The new city, which emulates China's wealthy coastal metropolitan cities, becomes an appealing site to accommodate those native residents who fled from their ancestral houses in the old town and now experience modern amenities such as air conditioning, automobiles, and broadways. After 2010, it was virtually impossible to find native residents who still lived in the old town but pursued no tourism business. Even the number of native business operators was in decline, partly because of rising rents that discouraged them from taking risks in the tourism industry and partly because of the town's highly competitive tourism market. Having said this, native residents have developed a deep sense of home in the town. As I further elaborate this sense, we can better understand their hesitation and struggle over whether to stay or move.

A True Love for Courtyard Houses

Native residents cultivate a sense of home from a true love for courtyard houses in the old town. The layout of a typical house has all rooms organized around a central courtyard. The courtyard is an important venue for socializing, relaxation, enjoying sunshine, and gazing at the clouds. The eaves provide shade while the residents talk and commune. For instance, Hexi (female, thirties) has already moved out to the new city, but her mother refused to do so. Hence, the family turned the spacious courtyard house into a guesthouse in which her mother takes care of the business. Meanwhile, Hexi visits her mother almost every day and handles important business matters. To Hexi, a yard is an essential component of a good living space. As she explains, "Naxi people feel quite uncomfortable if they cannot see the sun, and horrible if they cannot see the sky. We Naxi people highlight the importance of being close to nature. Otherwise we become disoriented." In the new city, she bought a house with a yard to maintain closeness with the natural environment. Similarly, Laoyu (male, fifties) explains the value of the ancestral courtyard house.

> Native residents of my age do not enjoy living in those apartments in gated communities. These apartments are made of concrete and steel. I feel very uncomfortable and suffocated when staying inside. Being used to courtyard houses, I don't like concrete apartments. These old houses are very nice, well ventilated, warm in the winter and cool in the summer. That's why I feel comfortable. If I move into an apartment, I cannot see anything once the door is closed. Here, I am close to flowers and water.

Both Hexi and Laoyu did not hide their fondness of courtyard houses, citing closeness to the natural environment as a primary reason. Indeed, the entire old town, together with courtyard houses, seems to be an exemplary imbrication of the natural and the human, enabling native residents to enjoy a homely life near nature without having to take adventures into mountains, jungles, or rivers. The association of a homely life with the natural environment suggests the importance of natural elements such as water, flowers, clouds, and sunshine in native residents' daily practices. This closeness can be achieved within courtyard houses, as described by Laoyu and others, to highlight a lifestyle full of warmth and comfort. Thus, a courtyard house is not merely a site for sleep and dwelling, but also an original ground from which they launch their whole life journey.

To Naxi people, a courtyard house is a central ground for daily activities and contributes to a feeling at home. Writing about settling down, Sanders (1993, pp. 121, original emphasis) asserts that "I cannot have a spiritual center without having a geographical one; I cannot live a grounded life without being grounded in a *place*." Connecting home to spiritual peace, Sanders (1993) emphasizes that a sense of belonging to a landscape creates a knitting between the self and the world. This connection can explain why native residents in Lijiang Old Town such as Laoyu and Hexi cherish their courtyard houses so much. To them, it is important to maintain their own culture – a culture of belonging rooted in the natural environment. They kept reiterating how comfortable and enjoyable the courtyard house was, spatially for its closeness to the natural environment and emotionally for memories from childhood.

Other native residents echo Laoyu and Hexi when highlighting their sense of home. Gengfu and his wife lived in a neighborhood along Wuyi Street. He inherited a large courtyard house from his grandparents and owned two small shops. One was rented out to a migrant for selling souvenirs, and the other was used by his wife to sell daily items such as snacks, beverages and drinks, and school supplies. When I met Gengfu for the first time in 2004, he told me that the number of native households in his neighborhood was 24. In 2007, the number dropped to four, and in 2010, two. Whenever I visited Lijiang, I would find Gengfu and chat with him about the town's transformation. His courtyard house was one of the best in the town (Figure 5.1). It has two storeys, is sunny and spacious, with a stream running in front of the fence. The courtyard was well maintained and decorated with flowers and bonsai. The couple continued to stay in this courtyard house because Gengfu's mother-in-law, the grandma in this family, refused to leave the town to live in the new city. In 2012, I could not find him anymore, as his courtyard house was converted into a splendid guesthouse and his two shops were used by migrants I did not recognize. His phone number had changed, and I lost contact with him.

Naxi people have a strong ritual of ancestor worship, which offers a mundane form of religion to ensure the family's cultural continuity. This continuity is spatialized in courtyard houses that were home to ancestors and inherited by their descendants. Through the house itself, Naxi native people can experience a family reunion with ancestors and practice a ceremony of worship to express gratitude for their protection. This gratitude is also expressed by a true respect to the living elderly, particularly grandparents in a house, as the most important persons in their family. In most cases, courtyard houses cannot be rented out if grandparents decide to stay inside.

Figure 5.1 The Courtyard in Gengfu's house in 2004. The lady in the picture was Gengfu's mother-in-law. **Source:** Xiaobo Su (Book Author).

A Close-knit Community in the Town

Beyond their own courtyard houses, native residents' sense of home draws on a close-knit community in which everyone knows each other, speaks Naxi dialectic, and celebrate common festivals. This kind of close-knit community is best described by Heyu, a retired resident in his seventies living in Gengfu's neighborhood.

> In the past, neighbors were like a big family. We got along well with each other. Kids ran from one house to another for candies and snacks. When a relative brought my parents some food and snacks from other places, every neighbor could share them. What a harmonious community! Now it is gone. Our neighbors come from other places for money. They develop only economic relations with native residents. When Gengfu's mother passed away, I went every day to help him. Why? I treated his mother as mine. When we were young, she cared for us like her own kids. This was a really harmonious society. What a sweet memory! At night, this memory becomes even more meaningful. Now we don't know each other along this street.

Harmonious neighborhood relations were cited many times by my respondents when they recalled how homely the old town was. This homely milieu in the town was constituted by casual encounter with friends, the prevalence of Naxi dialectic, and familiar streetscapes. For instance, Hexi still remembers a close relationship among individuals in one neighborhood. As she explains, "Naxi people are very hospitable. In the past, we constantly dropped by our neighbors and developed a strong bond among each other. I remember that during meals, kids brought their own bowl to visit different households and taste different foods. Our parents also got together to chat and relax together. Now my old neighbors moved to the new city and I never drop by my new neighbors who came from other places." Social relationships born out of childhood friendship further consolidated the town as a space of belonging. Native residents felt an attachment with everything in sight, including buildings, streets, canals, and bridges.

The old town in the past was both an expression of social familiarity and a spatial structure of collective community through bonds and interactions. Being a kinship society of its own, the town constituted a protected environment in which the arrival of strangers could be easily noticed with attention. Indeed, Lijiang Old Town before tourism development represented a typical form of Chinese traditional society, particularly in rural areas, characterized by familiarity, kinship, and geographical fixity. As Fei Xiaotong (1992) explains, people in traditional Chinese society see and interact with those who they have known since childhood. They do not have to select this kind of familiar society that they would live in, but rather are born into it. Hence, they develop a strong sense of familiarity with the society, as well as with the place, through frequent and repeated interactions over a long period of time. This sense symbolizes an intimate relationship, essential to a homelike community. Fei (1992, p. 41) explains:

> People in rural China know no other life than that dictated by their own parochialism. It is a society where people live from birth to death in the same place, and where people think that this is the normal way of life. Because everyone in a village lives like that, distinctive patterns of human relationships form. Every child grows up before everyone else's eyes, and in the child's eyes everyone and everything seem ordinary and habitual. This is a society without strangers, a society based totally on the familiar.

Fei goes further to explain that in a society characterized by this high level of familiarity, individuals can achieve a level of freedom whereby they can do whatever they please without worrying about whether they violate social norms. Embodied in rituals and customs, these social norms are integrated into the practices of everyday life so that following these norms is a natural thing in such a society. The old town in the past was a typical example of what Fei describes as *xiangtu* society, although those town residents did not rely on farming for a living. A sense of home among native residents in Lijiang Old Town was built upon inherent familiarity, down to the bone. From childhood on, they knew each other and played together, giving rise to an intimate community of care and strong bonds. By living in this sort of community, native residents felt secure and relaxed, which explains why some refused to move out to the new city, even though they lost such an intimate feeling of home after tourists and migrants occupied the town.

Community ties were further consolidated by a form of localized gathering. In Lijiang, Naxi people organize a regular gathering (called huacong in Naxi dialect, 划賨) to boost friendship and interaction. Originally, huacong was an informal financing mechanism drawing on rotating savings and credit among those who knew each other quite well, a mechanism popular in south and southwest China before 1949 (Tsai 2000). In Lijiang, women played an active role in organizing female-only huacong groups among classmates, neighbors, or simply, friends. Once the People's Republic of China was established in 1949, the local government in Lijiang enacted an order to regulate the organization of huacong. The various political campaigns against the evils of pre-1949 old Chinese society, together with widespread poverty, further dismantled huacong since women no longer had extra money to participate in informal savings programs. Since the mid-1990s, Naxi women have started to rekindle huacong as a form of socialization, different from various associations organized by Naxi men for tea or alcohol, so that they can have fun together and renew friendships (Zhao 2013). Nevertheless, its financial function largely disappeared, as people in Lijiang had more sources for monetary support. Meanwhile, Naxi men started to participate in huacong to interact with other members. Through huacong, local Naxi people build intense networks across different groups to maintain their social capital and economic ties. Migrants are rarely invited by local Naxi people to participate in huacong, which signifies the divide between these two groups of people (Chapter 3).

For native residents who still lived in the old town in the 2000s, huacong activities remained a key opportunity to renew friendship with their former neighbors who moved out to the new city. Hexun, a male in his forties, emphasized that "[i]nterpersonal relationships are worse than those twenty years ago. When I walk along this street, I cannot recognize anyone. They come to run a business or to tour the town. Here come eight persons. I don't know any of them." When his old friends moved out to the new city, Hexun still maintained some connections with them, particularly through huacong activities, to continue friendship, exchange ideas, and have fun together. As Hexun describes, "We met once per month. I knew where my friends lived in the new city. After a while, they might have moved to another place and did not tell me. I have no idea of where they are. It is not good to ask their location, which indicates that I might drop by for a visit. It is impolite." Once daily neighborhood interactions were replaced by sociable huacong activities, a sense of close-knit community disappeared.

The stark contrast between the old town and the new city further consolidates a sense of home among native residents, particularly among the elderly. Old Yang (male, in his late seventies) has lived in the old town ever since he was born. When I met him in 2007, he thought that the old town undertook a sea change in ten years and became less inhabitable. One of his sons already moved out to the new city and purchased a big house. It seemed that Old Yang and his wife were not persuaded to leave the old town, no matter how much their adult children described the advantage of living in the new city. For the couple, the old town was their home and they felt quite rooted and comfortable when staying in this specific homeplace. As Old Yang explains,

> We don't move, because my wife is unwilling to move. My younger son purchased a big three-story house in the new city. I think it is OK to live there, but my wife refuses to do so. Why? She says that the big house is like a prison. Our old courtyard house in the town is way more comfortable. Young people in my family move to the new city, and we two elderly stay in the town. When you walk around in the town, you can find that many dress in Naxi-style costume, but cannot speak Naxi dialectic.

Similarly, Zhao (retired, in his seventies) lived in the old town with his wife. Their three sons bought apartments in the new city and left the old town. During the interview, he kept mentioning how good the old town was and listed several counterarguments about living in the new city:

> We do not like the new city. Rather, we love the old town. We love it very much. Our sons and their kids live in the new city and return to the ancestral house in the old town during weekends. We understand that young people prefer to live outside. They persuade us by saying that air in the new city is fresh while the old town is quite polluted. But we feel uneasy about the new city.... I have lived in the town for my whole life and have become quite familiar with it. My wife and I are old. Elderly people prefer to live in old houses. This house comes from our ancestors and we are not accustomed to the lifestyle in the new city.

Obviously, Zhao and his wife have refused to leave home and give up their ancestral house. This refusal can be explained partly by a conscious choice of keeping a distance from the new city, the material metaphor of modernity, and partly by their familiarity with the old town.

The life experience of bell hooks can help us further understand this refusal. After a thirty-year stay away from her home state, bell hooks (2009) eventually decided to return to Kentucky and settle down in a place familiar to her. As she explains, this resettlement is to help her to seek healing by returning to "the Kentucky hills of my childhood, to familiar rural landscapes" (hooks 2009, p. 49). If such a return can help individuals to heal a restless soul and rebuild a sense of home, elderly native residents in Lijiang Old Town still face the pressure to leave the courtyard houses as well as familiar townscapes to start a new life in an unfamiliar setting. To bell hooks (2009, p. 52), a return to the hills of childhood enables her to "keep a sublime hold on life," a return aiming to claim back her rooted home. To native residents in Lijiang, however, a departure from their rooted home implicates them into a permanent unhomely life journey in which they must struggle for a sense of home. This departure is particularly challenging to those elderly such as Zhao and his wife, as they reiterate the fondness of the old town and their dislike of the new city. As bell hooks (2009, p. 18) explains, leaving home can evoke "extreme feelings of abandonment and loss. It was like dying." While Zhao did not present such an extreme feeling as described by bell hooks, his earlier quote betokens a strong sense of home in the town and a fear of loss in the new city. Nevertheless, this strong sense has become uncanny and their fear almost comes true in the process of tourism development.

From Familiar to Strange: In-situ Displacement

China's great transformation does not merely occur in the structure of political economy handled by ruling elites in Beijing, but affects the lives of ordinary people in peripheral cities such as Lijiang. Dramatic changes have happened within human relations, within daily patterns of dwelling and socializing, and within the cities where daily routines are organized and endowed with meanings. The phantasm of modernization often relies on the power of economic exchange and political mobilization to push individuals toward the race for wealth and fame. Ever since the tourism development of the late 1990s, the old town, like other cities in the world, is characterized by an influx and outflux of strangers – tourists and migrants – and accordingly, native residents have developed an uncanny experience of in-situ displacement when the place that they are familiar with becomes unfamiliar and alienated. As a result, native residents who remained in the old town realized that it had evolved into a strange place that they could not recognize. In other words, these residents lived in the town, owned property, and had access to public facilities, but they were in-situ displaced by migrants and tourists once the town undertook a transition from a familiar hometown to a bustling tourist destination. Instead of exploring how the tourism industry shapes the old town (more details can be found in Su and Teo's 2009 book), here I pay attention to how native residents perceive the town's transformation from familiarity to strangeness.

In Lijiang, the unhomely unfolds when some native residents realize that their comfortable courtyard houses become uninhabitable due to noise and pollution. In the past, they could easily fall asleep and wake up naturally, but in the 2000s, noise coming from night clubs and tourists disrupted their sleep. As tourism development becomes the primary business to spatially and socially organize the town, native residents' hope for a quiet place to live is more like an illusion than a reality. While some tourists visit Lijiang for loafing and retreat (see Chapter 4), more want to have fun, possibly even in a lunatic way. As Heyu describes,

> Guesthouses are everywhere. They surround us. At night, tourists still speak very loudly and play music. They have fun, but we have trouble. It is not quiet anymore, and we cannot have a sound sleep. One day, we must move out. What can we

complain about these guesthouses? It is business, and migrants come for money. No one cares about our sleep.

In a similar vein, Gengfu noticed all the tourism-related negative consequences brought by tourism development and expressed a strong desire to leave, not because he disliked his courtyard house or the old town, but because he cannot endure these consequences anymore.

> If possible, we want to move out to the new city. The old town is not suitable for dwelling, mainly because of migrants and tourists. They come to the town and have a completely different routine from ours. The night bars are still doing business at 3 a.m. and the music noise is unbearable. I want to live a peaceful and quiet life as I did before, but now it is unlikely. The Heritage Authority has ordered bars to reduce the volume after 10:00 p.m. It is impossible. The disco is too loud to bear.

In their hometown, Heyu and Gengfu feel an inexorable sense of anxiety about the quality of life, to the point that they seem to be displaced, albeit not through spatial dislocation. This sense of displacement transforms their courtyard house from a familiar and comfortable home to a strange place from which they want to escape.

Beyond their private houses, the old town at large is undergoing dramatic transformation as well. Heyu noticed that more and more Han migrants live in the town, and accordingly, fewer and fewer native residents still stay inside. The town has become more crowded than ever, but he feels lonely when walking along the noisy streets. If Gengfu feels unhappy about the night noise from bars that disturbs his sleep, Heyu is concerned with the disappearance of a familiar atmosphere constituted by dialect, festivals, and customs. As a resident who was born and grew up in the town, he was familiar with everything. Nevertheless, he complains that he cannot easily find someone who speaks Naxi dialect in the old town. Rather, he must speak Mandarin, a language foreign to him, with migrants and tourists:

> People do not speak Naxi dialect in the town. I feel quite strange. You cannot bump into acquaintances anymore, as most native residents have moved to the new city. The whole atmosphere is gone. The magic of Naxi dialect in the town is gone. Now we must speak Mandarin in the town. It is not our mother tongue. It is not easy for me to use Mandarin to express something humorous or subtle. Naxi dialect is beautiful, humorous, and sweet.

Although Heyu's house is not close to the old town's center, migrants have approached him to rent the house. He was reluctant to do so, as he admitted that he had a strong sense of attachment to his courtyard house. This sense included feelings, habits, and even the air. Despite that the town had turned unfamiliar to him, his sense of attachment still lingered. Here is how he calculated the deal between profit and meaning in terms of renting his courtyard house to migrants.

> I don't want to rent my house to migrants. This is the house I have dwelled in ever since I was born. No matter how much money is given to me, I am reluctant to leave. I guess it is a sense of nostalgia for the house. From the perspective of economic return, it is not a good deal. If I rent it out, I can earn CNY 60 to 70,000 (~US$ 8,850 to 10,320) in one year.[1] Then I feel bad. Now I am old and retired. If I move to the new city, what can I do if I cannot fall asleep? I will become very unhappy and disappointed. So, I can get some money from house rent, but lose happiness. A bad deal, isn't it?

Having said this, Heyu knew that one day he would have to move out to the new city as long as he was still alive. He just did not want this day to come so early, earlier than his death, so that when he left the town he became uprooted forever. This sort of struggle between stay and departure is described by Heyu:

> Once we rent out this house, we can move to the new city. But I guess I would feel uncomfortable, since I could not get used to a new place, including the air and water. It might take time to embrace the new city. How wonderful the old town is! Apart from the disordered things, it is a perfect place to dwell. Look at my lovely courtyard house.

Because of the tourism-driven great transformation affecting what was once taken for granted as the root of social experience in Lijiang Old Town, native residents have become strangers in a place they call home. Accordingly, the dramatic trajectory in which yesterday's familiar land becomes today's bizarre site reinforces their uncanny sense of in-situ displacement. The issue is more complex than the neat process of demographic replacement. Instead, I want to read Heyu's story as a metaphor for the unsettling effect exercised by the race for wealth and the optimization of space in Chinese cities. The transformation of Heyu's neighborhood, along with the dismantling of his social ties with old friends, gets entangled with the reconstruction of urban spaces in both the old town and the new city in Lijiang, as well as on a national scale the rise of consumerism – tourism, recreation, shopping, and so on – which draws on China's amazing economic growth in the past four decades and reshapes the composition of cultural affiliation. Nevertheless, individuals cannot feel at home in their regular setting – big cities for migrants and tourists, and Lijiang Old Town for native residents. They must use their purchasing power to remake a new home elsewhere – the old town is now for migrants and tourists, while the new city is for native residents.

Unsurprisingly, Heyu would move to the new city and resettle in a gated community there not far from his ancestral house in the old town. How much he could get used to a modernized lifestyle is unknown. The new apartment, together with the new city, was regarded as an uninhabitable prison, and now accommodates an unhomely life. Ironically, the modernized lifestyle, materialized in well-designed apartments in Lijiang and elsewhere in China, embodies the symbol of wealth and growth. The macro reconfiguration of China's economic development goes hand in hand with the microgeographies of in-situ displacement that Heyu and other native residents experience in Lijiang Old Town.

Native residents' in-situ displacement was accompanied by the increasing depersonalization of social and economic relations built upon exchange and calculation for short-term return. If tourism development brought new opportunities to reify the commercial value of native residents' courtyard houses, as tourists and migrants replace them as the main users in the town, it also brings about the price tag of increasing estrangement and inevitable displacement. A conversation with Zhao illustrates such a sense of strangeness:

Author: Any impact on your life brought by tourism development?

Zhao: A lot of impacts for sure, including rising living cost, pollution, and noise. Those outsiders throw dirty water and trash into the rivers, which is insane.

Author: How about your neighbors?

Zhao: Many of them have already moved out to the new city. Our neighborhood was far away from the town center and tourism development does not bother us very much. Nevertheless, about half of native residents move out and rent their houses to migrants for tourism business. Old friends keep moving out and we still stay in the town. If all of my old friends walk away, then the town becomes meaningless. Here it was quiet at night, but in Xinhua Street, the noise is so loud at night that native residents there cannot bear it. These unknown people sing, cry, and fight at night in bars. These bars are strange places with various dirty transactions. We are completely unaccustomed to them. They are not part of the old town. In the past, we knew each other in the town and everyone cherished honesty and kept their word. Now people cheat each other. The town is quite different, and cheaters are everywhere.

While individuals in China's coastal cities want to live in a condition of intense anonymity in gated communities, and no one holds themselves accountable to each other, native residents such as Zhao resist such a condition and cherish a sense of community. It is an unruly experience to encounter new neighbors every day, and never build a close relationship or engage in community building. Though migrants and tourists from other Chinese cities get used to a sort of pseudo intimacy in Lijiang, Zhao ridicules this pseudo relationship and recalls the prevalence of accountability among Naxi residents in the town.

Trust in China's rural society derives from familiarity, crystalized in long-term interactions, and rooted in customary norms. But trust in modern society draws on law and contracts on written protocols. As Fei Xiaotong (1992, p. 44) observes

China's transformation over seven decades, "created by strangers, modern society cannot incorporate the customary basis of rural society. Rejecting the customary ways of rural life, modern people denigrate everything rural. The rural village is no longer a place to which successful people want to return." In this context, Fei argues that we are forced to adapt to the modern society constituted by strangers whose interactions depend on written contracts instead of oral commitment. Lijiang Old Town represents one example of such a transformation that Fei describes in the context of China.

Since 2010, the old town has no longer been organized into a localized and familiar estate with strong bonds among its residents, but has revolved around the rational logic of commercial exchange for the sake of profit. Even the making of home for tourists by many guesthouse operators is largely a form of business strategy. Old social bonds are eroded and eradicated by exchange relations that are geographically promiscuous, as tourists and migrants come from a distance in order to meet in person in Lijiang for certain purposes. The erosion does not necessarily mean a permanent loss of housing property in the old town, since native residents still own these courtyard houses, not as a space of dwelling for themselves, but as a product of commercial exchange. As McQuire (2008, pp. 24, original emphasis) argues, the making and remaking of home in modernity is "the product of *both* the loss of stable coordinates, *and* the invention of new continuities and new processes of cultural affiliation across interlinked domains." Following this point, I argue that the town, like many others in China, undertakes a typical transformation from a *xiangtu* society characterized by familiarity and attachment to an alienated society characterized by estrangement and anonymity.

Home, for these native residents in the old town, is located in a selectively recalled and distilled past (McCarthy 2018). Their narrative of the past before tourism development represents a nostalgic yearning, always sweet, but forever gone. This yearning is akin to the challenge that the exile encounters: "of being displaced and yet capable of remembering place .... We have a great yearning, but we cannot fulfil it with anything but memory" (King 2008, p. 71). While these native residents still control their courtyard houses, the town itself becomes the source of estrangement and unsettlement when their routine life has been disrupted and their social bonds dismantled. When old neighbors leave the town, native residents must live with strangeness in two aspects. First, they encounter unfamiliar persons during their daily life and second, they are alienated by tourism-oriented activities in the town if they do not operate business. In Bauman's (1997, p. 30) words: "the question is no longer how to get rid of the strangers and the stranger once and for all, or declare human variety but a momentary inconvenience, but how to live with alterity—daily and permanently." While Bauman talks about liquid society in general, native residents' experience of living with alterity seems uncanny, as they endure the loss of home in their own hometown. Such a living thus conveys a sense of insecurity and anxiety. If an ideal home can foster familiarity and comfort in order to develop "the sensory experience of feeling at home that defies the insecurity and uncertainty that can come when confronted with difference" (Butcher 2010, p. 34), then the old town stopped offering such a sensory experience to native residents.

Going farther, we can regard sociospatial changes in the town as a form of uncanniness. In his essay "The 'Uncanny,'" which was originally published in 1919, Freud (1955) discussed the origin of the German word *unheimlich*, which is often translated into English as "uncanny," but more accurately, pertains to unhomely, a word less popular in the literature. To Freud, an uncanny sense could attribute not to what is strange or alien, but to the transformation of the known and the familiar into the strange, largely deduced by dread or horror. The horror, being social, emotional, or symbolic, can contribute to the destabilization and even dismantlement of the home for those vulnerable to external interference. The uncanny is "the fundamental propensity of the familiar to turn on its owners, suddenly to become defamiliarized, derealized, as if in a dream" (Vidler 1992, p. 7). Different from other scholars who separate the uncanny from the mundane, Freud (1955, p. 220) reiterates that "the uncanny is that class of the frightening which leads back to what is known of old and long familiar." In this sense, we cannot treat the uncanny as an experience foreign to individuals or reinforce familiar versus strange. The uncanny is frightening not because it is unfamiliar or new, but because what used to be familiar and old has somehow become strange and even alienated. The familiar-to-strange continuum characterizes a mundane aspect of China's great transformation in which individuals are thrown into an uncanny condition of struggling for a sense of home in their hometown.

The Freudian idea of uncanniness, that is, the familiar turns into the unfamiliar, is useful here when thinking about native residents' nostalgia for the old good days in the town. Their uncanny feeling about the old town is triggered by a transition from the familiar to the strange, by the blurring boundary between the public town and the private house. In Lijiang, the horror is driven by rapid tourism development and the large inflows of tourists and migrants. The homely, which contains familiarity, warmth, and established behavioral norms, dramatically becomes strange when the town is suffused with migrants and tourists who cannot be easily recognized. Regardless of their age, native residents' bittersweet and nostalgic narratives of the homely town suggest that they locate a sense of rooted home in the past, a sense which is distant and irretrievable, only existing in memories. When the loss of home feelings descends upon them, native residents make different choices.

Age Difference: Departure or Stay

Under the conditions of booming tourism development and large flows of tourists and migrants, the old town and its courtyard houses were overwhelmed with noise, congestion, tourist gazing, and finally, every form of materialistic temptation. The tourism industry produced a horror-like transformation, so conspicuous and irreversible that native residents have had no choice but to forsake their courtyard houses in exchange for monetary return. The elderly I interviewed in the 2000s persisted in residing in their courtyard houses for peace and comfort; but the young left the old town and chose to live in the new city.

Here I would like to narrate the struggle between departure and stay within one family to illustrate age difference. I met Laoyu in 2005 when I searched for nonstandard guesthouses. In the early 2000s, increasing numbers of guesthouse operators renovated courtyard houses by adding standard bedrooms, which means constructing restrooms within courtyard houses and thus completely altering their internal structure. Eventually, I found Laoyu's guesthouse and noticed he still maintained its original structure. Laoyu (male, late fifties) entered into the hospitality business in 2000. The goal was not to make a profit, but to protect the courtyard house. Laoyu and his wife lived in the house, while Deyuan, their son, stayed in the new city. Over ten bedrooms were vacant and Laoyu turned them into dormitories. Up until 2012, they refused to rent the house to migrants for hospitality business. Both Laoyu and his wife were retired workers with pensions. They did not rely on tourist money for a living. Rather, tourist money was used to repair and maintain the aging house. The house was one of the largest properties in the town, with two big courtyards and two small courtyards, well maintained with an "authentic" atmosphere for Laoyu and his wife to enjoy.

Before 2012, I chose to stay in Laoyu's guesthouse when I conducted fieldwork in Lijiang. Rather than undergoing substantial decoration and improvement, the house remained in a very mundane status (see Figure 5.2) and none of the bedrooms had attached toilets, big-screen TV, or air conditioning. By 2010, tourists were not interested in staying in such relatively modest living conditions. I enjoyed it, not so much to save money, but to experience the lifestyle of native residents. When I went to Lijiang in 2010, I certainly understood that this could be the last chance for me to experience such a modest, or traditional, living condition, before courtyard houses in the old town became modernized or reconstructed for tourist business. During my two-week stay, very few tourists even bothered to ask for the availability of bedrooms in Laoyu's guesthouse. By 2012, the couple moved out to the new city and gave up the right to use it for daily living, and I had no further chance to meet Laoyu and his son Deyuan (a subject to be discussed later).

Laoyu and his wife lived on the first floor on the left, and other rooms were converted into bedrooms (Figure 5.2). I normally stayed in a bedroom on the second floor on the left. When I completed a tedious day of fieldwork, I sat on the couch in the left corridor to enjoy the warm sunshine or chat with Laoyu.

I met Deyuan in the guesthouse when he came to visit his father Laoyu. He was surprised to see that I stayed in the front courtyard to taste tea and look at flowing clouds. Although I cannot speak Naxi dialect, I frequently chatted with Laoyu using Yunnan dialect. To Deyuan, I was a strange guy, more localized than native residents enjoying the town. Then he joined me in the courtyard and

Figure 5.2 The front courtyard in Laoyu's house in 2007. **Source:** Xiaobo Su (Book Author).

chatted about the old town. As Deyuan explained, the town was the place where he was born and grew up, but has become an unfamiliar place:

> The old town has evolved into a different place. It is not guided by native residents, but by migrants and tourists. It is no longer a place for natives. We rarely enter the old town. Because my parents still live inside, I must visit it. To be frank, I haven't dropped by Sifang Square in about ten years. It is not necessary. The town is filled with entertainment and consumption, giving rise to an inhumane existence that I cannot recognize. It is a site of consumption for profit making, or for the sake of so-called GDP. More and more big corporations started to invest in the tourism industry in the town. The business is not to promote Naxi ethnic culture, but to commercialize whatever is profitable to make money.

Deyuan lived a frugal life alone in the new city and visited his parents in the town once a week. After reading traditional Chinese scripts, he felt quite disappointed at the dramatic change in the town and chose to move out, particularly after his previous tourist business became an abject failure. To him, the small apartment where he lived in the new city was his real home, where he could continue his wood carving projects to sell to tourists in a friend's shop in the old town, but avoid endless interruption by curious tourists. Asked where he called home,

Deyuan made it clear: "The town is my hometown, but I do not have a strong sense of home anymore." It seems that his every visit home left him torn: he needed to escape from the town, but he had to visit it because of his parents. Being unable to feel at home in his hometown, Deyuan engaged in an endless run away from home, not because of domestic violence or precarious housing, but because of a hatred for the ills brought by tourism development.

For young native residents such as Deyuan who have left their homes in the old town, they disconnect from a personal history embedded in social and spatial relations and prefer to evade in-situ displacement to such an extent that they have to bear the loss of home feelings. For those natives who leave the town, this loss is more emotional than materialistic, since they still own the courtyard houses and receive rent to share the economic benefit brought by tourism development. Nevertheless, the courtyard houses, no longer used by them, become a space of consumption for lifestyle migrants and tourists from big cities. As Deyuan noted to me, his parents had to move to the new city sooner or later. After all, the courtyard house was economically valuable in the tourism market. Their departure came in 2012, and since then, I have never had a chance to live in a modest guesthouse with a Naxi family in the old town.

By 2010, many young native residents had already moved out to the new city either because of unhappiness with the town's transition or because of modern lifestyles in the new city. Hexi, another respondent who had already left the old town, explains the perceivable changes to herself, the neighborhood community, and the town as well:

> Now we live quite far away from each other. Of course, we participate in different huacong groups once per month and gather in a farmhouse in Lijiang's vicinity. I hang out with friends every Saturday and Sunday. But I don't know where my old neighbors live now, even though I can meet some of them in some huacong activities. They move to the new city and in the long run, the bond falls apart. After all, it is not easy to resist the temptations brought by money. Facilities in the new city are better than those in the old town. It is quite inconvenient to live inside. Elderly people prefer the old town, as they can interact with their lifelong friends. Sometimes young people feel quite constrained. That is why I moved out to the new city.

Hexi regularly participates in huacong activities to maintain connections with old friends, but admits that it is impossible to recreate those reciprocal neighborhood relations. Furthermore, like retreat tourists and lifestyle migrants who must handle the deal of modernity, Hexi and other natives cannot resist the lure of financial gain brought by courtyard house rents.

For those young residents still living in the town, their desire to move to the new city is as strong as for those who left. Different from the elderly who lingered in the town after their retirement, young native residents can bid farewell once they can pay for an apartment in the new city. Hexiao (male, in his thirties) is among very few young native residents to live in the old town. He has two storefronts and rents them out to migrants for good return. Although he dreams of being his own boss, he realizes that the business pressure could easily overburden him. Hence, he chooses a safe way by renting the shops out to migrants. He still lives in the town and enjoys himself very much. During the interview, he admits that he will move out to the new city once he has adequate money to buy an apartment. Reflecting on the town as home, Hexiao notes that:

> Native residents of my generation still interact with each other. As long as my neighbors have important events such as weddings and funerals, I must help them out. This is a good interpersonal relation. We are born in the town and grow up together. That is why we have a strong sense of bond. When younger generations move out to the new city and have no interaction with each other, this kind of relation automatically disappears.

The difference between the younger and older generations is quite explicit. The elderly attempted to defend a home where they could fall asleep calmly and maintain an organic relationship with the town. Most of the remaining native residents that I interviewed during my fieldwork in the 2000s were retired, still active, and independent with good health. They relied on pensions instead of tourist money for a living. Their affection for the town was so deeply ingrained that it was not easy to leave their ancestral houses to migrants, although their adult children had already moved out to the new city and might keep persuading them to move out. They must spend time, energy, and resources maintaining their houses and endure all the negative impacts brought by tourism development. The longer they live, the more they cherish their ancestral house, a property that they still control and inhabit, even though the town has become a playground for tourists and migrants. Once they become dependent on their children or cannot tolerate the noise or congestion anymore, they have no choice but to give up their right to use the ancestral house and bid it farewell. While the elderly are nostalgic for the old good days, the younger generation is more realistic and forward looking. They firmly embrace modern facilities in the new city and enjoy a lifestyle akin to those in China's well-developed coastal cities. In this sense, age difference, often associated with and exemplified by the choice between stay or departure, is performed and reinforced in the process of homemaking and unmaking.

To some extent, Lijiang's new city that surrounds the old town is developed with concrete buildings for the sake of economic development (Figure 5.3). It becomes a clumsy mimicry of China's coastal cities, emerging as the representation of modernization that is itself a process of creative destruction that subordinates individuals to political and economic power. While Bhabha (1984) emphasizes that mimicry aims to create a reformed, recognizable Other in the context of colonialism, I hesitate to repeat the notion of internal

Figure 5.3 Streetscape in Lijiang's new city in 2007. **Source:** Xiaobo Su (Book Author).

colonialism, as used by scholars such as Oakes (1998) and Chuluu (2021), to frame the relationship between Lijiang and coastal cities. Rather I treat mimicry as a powerful instrument of globalization and modernization to homogenize urban development that adheres to the dominant function of real estate and mass consumption in shaping Chinese cities ranging from the coast to the hinterland. To Bhabha (1984, p. 126), mimicry can generate "a subject of a difference that is almost the same, but not quite." It is this same, but not quite, condition that colonial entities fail to replicate their European benchmarks, but end with an in-between journey – unreturnable pre-colonial societies and unreachable European modernization. By all means, Lijiang's new city never becomes another Shanghai or Shenzhen, but mimicry fulfills the pent-up desire among Lijiang locals for modern lifestyles and thus facilitates the circulation of capital between the new city and the old town. While those in coastal cities can travel to Lijiang for a sense of home and thus pursue a moment of loafing and calmness (see Chapters 3 and 4), locals have no way to escape a similar specter of homelessness born out of mimicry. Thus, they engage more frequently in shopping and social gatherings such as huacong so that they play mahjong, drink, and kill time.

This kind of age difference occurs in other Chinese cities. In Beijing, while middle-aged adults have moved swiftly out of the old city and live in high-rise apartments equipped with modern amenities, their aging parents get stuck in dilapidated courtyard houses and face various challenges in their daily life. Having said this, these Old Beijing elderly residents are familiar with their surroundings and the neighborhood, and enjoy the biographies and memories that their houses carry within and around them (Nguyen 2017). To these residents, their residential blocks are not a capital-land of urban redevelopment and profit making, but represent a homeland embedded in lifetime histories and personal trajectories. Once their neighborhoods are replaced by high-rise apartments, shopping malls, and fashion shops, they are spatially displaced into a condition of homelessness simply because their homeland and related vernacular pasts are gone. To defend their social and economic interest, some elderly residents in Old Beijing mobilize site fights to slow down and even disrupt redevelopment projects in their neighborhoods. Waiting, or the protraction of time, is deployed as a temporal means of subtle resistance to belie the linear, speed-oriented logic of construction and development in Old Beijing.

In Lijiang, such a resistance rarely happens, as natives are not forced to leave. Once the familiar town becomes unfamiliar, the majority of residents choose to resettle in the new city, while others stay but have to endure in-situ displacement. Regardless of their choice, they enter an unhomely condition, as Bhabha (1994) elaborates, in which the practices of everyday life are disoriented. Foregrounding the struggle between staying in the town and moving to the new city offers a way of reading emergent signs of the contradiction between modernity and tradition, and money and meaning. To understand this struggle, it is necessary to add a critical dimension of political economy to Bhabha's usage, in order to interpret the formation of unhomely life generated in the process of a local but major transformation as symptoms of modernization at the periphery of China. Steady economic return brought by house rent, plus inconvenience brought by heritage protection and tourism development in the town, incites most native residents to move to the new city so that the town becomes a new home to migrants and tourists. Having said this, the stark contrast between the elderly and the young reminds us of the complexity of in-situ displacement and highlights different temporalities – slow versus rapid, old versus young, traditional versus modern – which are actively obscured beneath the process of making and remaking home in different geographical contexts.

Making and Remaking Home in Daily Life: Four Stories

Analyzing native residents' efforts to make and remake home allows us to understand individuals' compliance with and struggle against speedy modernity in the context of China's great transformation. Home can be uprooted, not in a fixed site but through a different experience of time and space. In the following, I narrate four stories to unravel the complicated contour of homemaking by native residents to negotiate the deal of modernity.

Story 1: Making a Home for Tourists

The first story is about Huai (female, in her sixties) and her husband Xianghe. In 2004, I met them in their courtyard house where they operated a hospitality business for tourists. The couple did not upgrade to standard rooms, since they refused to change the house's original aura and structure. Their children lived in Kunming. In addition, Xianghe's mother also lived in the courtyard house. The selling point of their guesthouse was not facilities, decoration, or free fruit, but a sense of home embedded in ordinary daily activities in a typical Naxi family. As Huai explains,

> Some tourists also enjoy our guesthouse, although we don't have standard rooms. They can experience a different sense, a sense of harmonious family and daily peace. My husband and I live here too so that our customers can talk to us, pretty much like a family .... When doing business, we care about integrity. We treat young customers as our kids and middle-age ones as our friends. I hope to get along well with them like family members. This is our own property and there is no rent. Both of us have pensions for our livelihood and thus face no economic pressure. That is why we have a good mood. There is no need for us to make fast money.

This sense of home is recognized by Xianghe who saw the need to offer a home away from home for tourist customers.

> When tourists come to stay in my house, they can feel at home. First of all, actual Naxi people live here as a real family. Second, tourists are welcome to chat with us and exchange some ideas. If they have trouble, we can help them out. We locals have resources to solve their problems. I don't charge extra money .... I think that a sense of home is really important to tourists. When people visit an unfamiliar place, they definitely want to live in a comfortable site. Otherwise, they spend money for trouble. That is weird.

To Xianghe, a home for tourists means comfort, zero trouble, and relaxation.

Both Huai and Xianghe were retired, and their children were not around. It seemed that they need not look after grandchildren. To them, running a guesthouse was more like a social activity to interact with tourists than an economic business for profit. Interestingly, both enjoyed social and economic relations with tourists.

> I feel a little lost if no tourists stay in my guesthouse. When they stay, I can talk to them and learn some things. My own children are living in Kunming and keep busy

> with their own work. They don't frequently return to Lijiang to stay with us. To provide service to tourists, both my husband and I have some things to do so that we don't feel so bored. Life has more fun. (Huai)

> I want to run this guesthouse because I obtain fun from it. Now I am retired, and have time to run the business. Otherwise I have nothing to do. In addition, tourists tell me something I don't know, and I share my knowledge of Naxi culture with them. Eventually we become friends. That is really a fun experience. (Xianghe)

When I asked them whether they planned to rent their courtyard house to migrants and move to the new city, their answer was an absolute no. As they explained, they faced no economic pressure, nor were they forced by someone else to move out.

> We don't want to rent it out. Even though we would close the hospitality business one day, we will stay here on our own. Anyway, we are not short of money, as we have a pension. Nevertheless, the business gives my life more fun. (Huai)

> To be frank, I cannot allow migrants to rent my house for business. They are not going to preserve this house, but radically change it for money. For sure we want to make some money, but we don't rely on tourist money for a living. This is a different mood. (Xianghe)

The story of Huai and Xianghe does not merely highlight home as "a sense of cultural belonging and existential shelter," but also specifies the ways in which individuals can retrieve the meanings of life in the process of making home (McQuire 2008, p. 8). If modernity is a deal of trading meaning for money, then making home for themselves and tourists in the town enables Huai and Xianghe to renegotiate this deal by prioritizing meaning over money so that they can maintain a sense of home, and thus feel comfortable, in their daily life. The loss of home, which Berman (1982) posted as the fundamental condition of modernity, can be at least understood as the dominance of money over meaning. This dominance, ever since the Industrial Revolution, exerts a profound impact on the way in which individuals define and experience home in the present, whether at the level of daily dwelling, commercial city, or the national project of modernization. Even for those living in Lijiang Old Town, they have to figure out what to choose between profit and meaning.

After careful calculation, most town residents seek economic value from renting their houses out to migrant businesspersons rather than from directly operating tourism businesses in the town. Asked why they willingly become rent seekers, most of respondents replied that they cannot compete with migrant businesspersons in terms of investment, management skill, and innovative business ideas. Sometimes this kind of comparison can easily evolve into ethnic difference between Han and

Naxi people. While Han people are regarded as more entrepreneurial and skillful, Naxi respondents constantly admitted that they cannot compete with Han migrants in the tourism market. Rather, they choose the safest approach to share tourism benefits rather than take risk in a highly competitive market. Instead of renting their house to migrants, Huai and Xianghe chose to run the guesthouse on their own, for the main purpose of creating a sensed home for tourists so that they could have genuine interactions with their customers. While both understood the popular presumption that Naxi people cannot be compared to Han migrants in terms of business operation, they did not believe in such a discourse of ethnic difference. Rather, they harnessed their advantages, including little economic pressure and Naxi lifestyle, to operate their business.

Story 2: Promoting Naxi Culture for Profit

The second story is about Yumei, a Naxi native resident in the town. In her forties when I met her in 2005, Yumei was one of the very few native residents who has succeeded in running a high-end hospitality business in the town. She graduated with a master's degree in cultural anthropology from a university in Kunming and became a lecturer at a different university. Later, she quit the job to become a tour guide for international tourists. This work experience enabled her to understand how the tourism market operates. Seeing Lijiang's booming tourism industry, she returned to her hometown, found a job in a starred hotel in the new city, and then got promoted to be the general manager in the hotel. Again, she quit the job and started to run her own high-end guesthouse in the town. It had sixteen standard rooms, with two courtyards, and its room rate in 2004 was CNY 360 (~US$ 53.1) per night, in a stark contrast with CNY 100–150 (~US$ 14.75–22.12) per night in other guesthouses in the town. In 2021, Yumei's guesthouse remained a benchmark in the town, with a room rate of US$ 180–200 per night, on par with the Banyan Tree resort in Lijiang.

Yumei had a very clear vision about how to run her guesthouse. Like Huai and Xianghe, Yumei also regarded a sense of home as a key selling point to attract tourists, but she deployed a different manner to achieve the goal. As she explained,

> I want to foster a sense of home among our customers. This sense is different from free access to a house, although sometimes we think free entry and exit is a precondition of my home. It is more like a place for restless travelers to find peace, pretty much like a return to one's hometown. Tourists can find sustenance in our guesthouse .... People in big cities have already left their original hometown, and might feel uprooted. When they come to stay in our guesthouse, they can retrieve some feelings that they may have toward their hometown. A place can be called home if it can help us to rest our soul. It is not merely about birth or dwelling. Lijiang is a home to many people. They can be themselves at will.

In her guesthouse, Yumei highlights a sense of home for tourists entails two aspects: high-quality homely service so that tourists need not worry about domestic chores, and a home of Naxi culture to enable tourists to develop a better understanding of the place. The selling of home implies the operation of experience economy, an area where home and profit can combine and transform into each other in Yumei's business. Inevitably, home is endowed with commercial value. The synthesis of home and money means that home can be regarded as "an economic space, a breeding ground for consumer fantasy and a site of consumption and consumer display" (Osborne 2000, p. 53). To Di Domenico and Lynch (2007), the production and consumption of such a hospitality draws on culturally intangible connotations of the home to transcend standardized hotel service for more unique experiences. Like those lifestyle migrants who run guesthouses as a home away from home, Yumei turns her guesthouse into a home-related business, via an experience of homely life, to enable the logic of capital accumulation to reach deep into the intimate sphere in which tourists are supposed to feel at ease.

As a native Naxi, Yumei felt obligated to promote Naxi culture and realized the business potential brought by this promotion. She envisioned that her guesthouse was a platform to target tourists who wanted to spend money for cultural experience. Therefore, she organized a party once a month to invite customers to taste local tea, snacks, and alcohol made in Lijiang. Most of her customers were foreigners. She took the opportunity to spend thirty to forty minutes explaining Naxi culture and Dongba religion. This is a value-added service that allows customers to socialize with each other and develop more understanding of the local culture. As Yumei explains,

> I use the guesthouse to run the hospitality business and promote our local culture. This is what I really want to do. For instance, customers need to pay for the tea parties. Indeed, these parities allow me to learn from my customers and meanwhile, make money. I am very happy to narrate the cultural essence of our Naxi people. My opening remark starts with "I am a Chinese woman" and then "I am a Naxi woman." I want them to understand that the party takes place in China and is managed by a woman with an ethnic minority background. I never feel inferior as being a Chinese, a woman, or a Naxi person. Only if I love my own ethnic identity can I love China. I want my customers to get this point.

This utterance demonstrates that Yumei presents a different mentality by taking a risk to run high-end guesthouses in the old town. As a female entrepreneur, Yumei hopes to verify that in Lijiang's highly competitive tourism market, women can succeed in the tourism business. By standing up for female Naxi entrepreneurs, Yumei represents an awareness of ethnic and gender equality in comparison with the dominant groups.

Furthermore, through her life journey from Lijiang to Kunming and back to Lijiang, Yumei displays a cultural competence in local Naxi elements and being a Naxi woman. Her entrepreneurial skill to bring these competencies to a niche market reinforces her connection with the town. Her comments on creating a cultural oasis for tourists also point to the embodied nature of homely hospitality, that is, comfort and relaxation in a cultural milieu free of daily chores and anxieties. This story foregrounds some of the themes in my analysis of homemaking in the town by some natives, including the appeal of running tourism businesses in the town, the performance of Naxi culture for tourists, and the ability to manage their life journey in relation to a changing world.

Why do ethnic entrepreneurs such as Yumei emphasize that it is their responsibility to promote and defend Naxi ethnic culture? Their successful tourism businesses ensure that local people are as good as migrants in terms of business management. Through her business, Yumei attempts to reject both ethnic difference between Han and Naxi and gender difference, as she confidently proclaims her identity as a Naxi female in Lijiang Old Town. This reflection on stereotyped differences incentivizes Yumei to better understand Lijiang as her home in relation to her customers. Thus, home is not located elsewhere, but deeply embedded in her mind. This embeddedness is flared and reified into a form of home experience to characterize Yumei's hospitality business. In addition, migrant Han businesspersons encroach on the town and become the dominant actors in Lijiang's tourism market. Many businesspersons give priority to profit and take every chance to capture tourist revenues. Thus, to ethnic entrepreneurs, the tourism market becomes a "battlefield" in which they use their economic wisdom to defend Naxi culture. This responsibility entails "the release of noncapitalist imaginaries into the constitution of economies and economic structure, and the defense of local cultures from their normalization by dominant cultures" (Escobar 2001, p. 164). Nevertheless, in Lijiang, this release can be made possible only by capable entrepreneurial skills in the tourism industry.

Story 3: The Life Cycle of a New Lijiang Local

The main characters in this story and the next are a couple, but I narrate their stories separately to unravel how home is made and remade in Lijiang. Yongxin travelled to Lijiang in 2005 to meet his pen friend, Haitun (the next story's main figure) who was a native resident in the town. To continue their romance, Yongxin annually travelled to Lijiang to meet Haitun and stay for a while. After graduating from a university in 2007 in Jiangsu province, Yongxin went to Wuxi, a rich city in the same province, and found a job in a subsidiary of a Fortune 500 corporation. In 2011, he quit his job and migrated to Lijiang, and in 2015, married Haitun. When I met him for the first time in 2018, Yongxin and Haitun had one son and one daughter. He also brought his parents from Zhejiang province to Lijiang and purchased a spacious house in the new city, not far away from the courtyard house where Haitun's parents still live in the old town. Yongxin has become a new Lijiang local.

In the city of Wuxi, Yongxin had a successful career. In two years, he became a sales manager. Nevertheless, he started to realize that he did not feel well due to the extremely busy pace of travelling in different cities. He suspected that he suffered from depression, but could not tell if that was the problem. He underwent several medical examinations in well-known hospitals in Shanghai, but was told that his body was healthy and there was no medical problem. Even though he was busy, he insisted on reading books, which enabled Yongxin to develop a critical understanding of the meanings of life. Haunted by depression, Yongxin decided in 2011 to pull himself out of the hectic race for wealth and stay in Lijiang to reset his life. After a two-year-long adjustment, his sense of depression naturally disappeared in 2013 and he felt normal again. In 2018, he described how he spends a regular day:

> My life is very simple. I wake up naturally and prepare for breakfast. When is that? I neither know nor care. When to wake up or go to sleep depends on the natural need of my body. I just follow the cycle. After breakfast, I spend time watering flowers in the yard and then read books. Around 11:00 a.m., I go to the café shop and help Haitun until 10:00 p.m. It is not busy at all. I keep reading books in the shop.

This sort of simple life in 2018 was completely different from what Yongxin experienced before migration. Yongxin's life journey from Wuxi to Lijiang can be regarded as an antidote to and even a resistance to "the fast paced imperatives of global capitalism that urge the entrepreneurial self to speed up, become mobile and work harder in order to be valued as successful, productive and conspicuous consumers" (Fullagar, Markwell, and Wilson 2012, p. 1). By engaging with slowness, people can step off the treadmill or refuse the dominant logic of speed. To Yongxin, changing the imperatives of global capitalism could be too much, but an escape from them is possible, partly because he accumulated adequate savings through his deep embrace of time discipline and partly because he chose to embrace Lijiang lifestyle.

In May 2019, Yongxin started to make some changes. As he explained, he started a business in February 2019 after several years of preparation, in order to handle the economic pressure once his second child was born. The need to secure a decent standard of life for his family pushed him to become an entrepreneur, and his business mainly focuses on the sale of Lijiang's local cultural products to customers in big cities in China. Yongxin has confidence in his entrepreneurial skills. Once the business was launched, a sea change occurred:

> My life becomes quite busy, and the transformation is from a Lijiang style to a Shanghai style. What a difference! Now I don't have any spare time. No time to

enjoy life, and my day revolves around the new firm. The business is quite good and I have to handle the pressure of production.

Before he started the business, he had plenty of time to read books and drink tea either alone or with friends. In 2019, he had no mood to brew a bowl of tea. As he said, "I have no time to drink tea. Water is fine." To him, the value of water has changed. In the past, he spent much time watching the constant flow of water in the town's rivers and boiling water for tea, which allowed him to calm down and stay away from busyness. Drinking tea was a form of spiritual practice through which he could keep a distance from the shadow of depression. Now he drinks water for basic need, as he has no time to make tea. Accordingly, water becomes a source to meet basic need, instead of a medium for spiritual reflection. Resuming the cult of speed, Yongxin changes his idea of time: "Time means value. The most valuable thing I can do is to get what I want efficiently." To Yongxin, busyness or loafing are both life choices to normalize an appropriate temporal order compatible with the changing relationship between himself and the world.

In October 2020, Yongxin took part in our third interview in a day before he travelled to meet a venture investor in Shenzhen. His business had been rapidly expanding into different fields and cities. As he explains, his firm attempts to establish a database for algorithm analysis so that consumers and producers can efficiently build connections and find what they need in the food market. Talking about his business ambition, Yongxin joked that he wanted to establish a Fortune 500 firm from scratch.

Yongxin: We want to build connections. The catering sector is our starting point and later on, we will expand the algorithm to other sectors and eventually build a business aircraft. To get bigger and stronger!

Author: It seems that you pulled yourself out of a Fortune 500 corporation, and now hope to build another Fortune 500 corporation.

Yongxin: You are right. Eventually we must do things. The long recuperation is not a waste, but helps me understand the meaning of life and establish a new home. It is an important experience, isn't it? Now is the right time to sail for a new journey.

Asked whether he wanted to migrate back to one coastal city so that he can further develop his business, Yongxin did not answer directly. Rather, he used the metaphor of water to emphasize the need to become resilient. As he said, water takes the shape of glass after being poured into a glass, and the shape of bowl in a bowl. Indeed, he mentioned "flow" many times during interviews, which indicates the value of resilience in and adaptation to different contexts. In other words, he enjoyed a relaxed life and fixed his time-space in the old town during his eight-year odyssey. Now he enjoys a highly hectic life after starting a new business. As he points out, "Don't stick to a fixed label. Why do you place yourself in a rigid category? A label or a category is not important."

Yongxin's emphasis on flexibility and fluidity is an exemplar of what Bauman (2000) describes as liquid modernity. To Bauman (2000, p. 2), fluidity or liquidity is a fitting metaphor of modernity: "Liquids, unlike solids, cannot easily hold their shape. Fluids, so to speak, neither fix space nor bind time." Yongxin's preference with flowing water spells out the need to unfix space and unbind time in a highly competitive market. Thus, Yongxin disregards the idea of a house as essential to home, and emphasizes the need to think about home as socially constructed. As Yongxin describes,

> What does a house refer to? Is it a place for dwelling or an embodiment of meanings? To me, a house is a materialistic construct. You can redeem such a construct anywhere and own it. You cannot be limited by a house. What is home? It is where people are. Of course, a house or an apartment is the foundation of a home.

Having said this, a liquid man such as Yongxin should not be defined as houseless. "Liquid" individuals have adequate pecuniary ability to purchase property and enjoy a comfortable space of dwelling. Yongxin's reembrace of busyness appears to contradict the sense of homely life that he wants to pursue in Lijiang. Yet it offers a glimpse of how people engage with or keep distance from busyness in different conditions. As an entrepreneur based in Lijiang, Yongxin develops his business on the basis of big data and algorithm analysis, both of which are quite popular in China's first-tier cities such as Shenzhen, Shanghai, and Beijing, but have not activated new business models in peripheral cities. The success of his business brings both commercial return and political fame to him. In 2020, Yongxin was recognized by Lijiang Municipal Government as a high-level talent so that his firm can enjoy tax reduction and public subsidy. Being appreciated by the local government as a role model of new Lijiang residents, Yongxin's settlement in the town lays the foundation for his extra-local business expansion in other Chinese cities.

Here I follow Mallett (2004, p. 76) to stress that regardless of how home is conceptualized, it represents a spatial and relational realm from which "people venture into the world and to which they generally hope to return." From a journey-based perspective, home is a place of origin and a destination of return. Between the origin and the destination, individuals conduct one journey and another, across space and over time, so that they can live with this actual world. Yongxin's life cycle from busyness to loafing to busyness illustrates the spiral of unhomely life. Indeed, Yongxin admits that he has undertaken an odyssey.

> Yes, I am quite busy, as much as I was before I came to Lijiang in 2011. But I am willing to do so, since this is the business I love to operate. Willingness is very important. It is great if you can devote yourself to a significant and meaningful thing.

> According to Freud, life is an odyssey of self-discovery. We keep searching, but don't know what we like or what we want to do. What a painful journey! I think I have completed this journey after an eight-year recuperation [in Lijiang]. Now I feel quite OK with my current lifestyle.

Yongxin's story shows that home is not unmade, but constantly remade. Before coming to Lijiang in 2011, Yongxin suffered from a hectic lifestyle that deprived him of a sense of home. During seven years of recuperation for mental recovery, Yongxin incorporated simplicity and loafing into a homely way of life. Once he fully activated his business ambition in 2019, he defined home as where people are, a liquid space of social interaction. Yongxin's life journey is swinging between one home and another, showing an interesting twist regarding Lijiang: Yongxin as an insider resumes a busy lifestyle to promote his business from Lijiang to coastal cities in China, while those living with busyness in these cities travel to Lijiang for simplicity and loafing. In an endless search for home to withstand homelessness, Yongxin establishes his sociospatial relations with the world and projects his own cultural value upon different locales.

Story 4: Being at Home Forever

Haitun is one of very few young native residents who still spends much time in the old town. Her parents live in a courtyard house in Qiyi Street. When I met Haitun in 2018, she and her husband had purchased a big house in the new city, but she visits her parents almost daily. Inspired by the interesting experience of being a pen friend with Yongxin in the 2000s, Haitun persuaded her parents to convert the front rooms into a shop and started to run a postcard store to cater to tourists who wanted to send a written message from Lijiang to their friends. Without a need to pay rent, Haitun did not have much business pressure. Nevertheless, the business was not as good as she expected. Once Yongxin migrated to Lijiang in 2011 and established a stable romantic relationship with Haitun, they decided to switch to a café when domestic tourists got used to coffee and preferred a quiet place for loafing (see Chapter 4). While Yongxin paid more attention to recuperation, Haitun was the main operator. Rarely thinking about profit, Haitun enjoyed her business more like a social opportunity to interact with tourists from all over the country. As she explains, "no matter what business I did before, I didn't have much ambition. As long as I could meet different people in my shop, I felt very happy. That's enough. Everyone came here with stories and I loved to listen to their stories. Hence, I had nothing to worry about, but felt comfortable."

Different from other Naxi native residents who either felt reluctant to interact with tourists or carefully calculated the benefit and cost of running a tourism business, Haitun's laid-back business style is related to her personal experience. When she was very young, Haitun stayed with her mother who worked for a state-owned factory far away from Lijiang Old Town. At the age of two, Haitun was sent to the factory's childcare center. Later, her mother returned to the old town and Haitun went to a local primary school. Technically, part of Haitun's childhood was not spent in the town. Once she started to live in the town, she was treated by other children as an outsider and had a hard time to develop a close-knit community with her peers such as the one between Gengfu and Heyu. As Haitun admitted, "those children had their own friend circles. Of course, I made some new friends, but not very close. We could not cordially blend together, as we had no common topics." Since Yongxin moved to Lijiang, Haitun has had more opportunities to establish social ties with migrants and get exposed to the tourism market. She frankly admits her husband's explicit influence.

> I am more or less influenced by my husband. Nowadays I am not quite connected to the local Naxi community. In other words, I make friends with migrant entrepreneurs running guesthouses or gift shops in this street, or have a friendly conversation with tourists like you guys. I rarely have thoughtful conversations with local Naxi people, although I am a native Naxi resident in the town. One reason is that their lifestyle is quite different from mine. For instance, they go to karaoke shops at night or have huacong to play mahjong on the weekend. I have disliked this style ever since I was very young.

Since Naxi people rarely invite migrants to participate in huacong activities, migrants in the old town create their own community through regular social gatherings and unsurprisingly, exclude native residents from this community. Haitun seems unusual, as she keeps distance from Naxi lifestyle and maintains rapport with migrant entrepreneurs. In this regard, her sense of home is not built upon close-knit community ties as much as my other native respondents reiterated, but upon her subjective feeling of being comfortable in the old town.

As Haitun observes, the old town is technically occupied by migrants, and very few native residents still live inside. To promote tourism development and attract tourists, some Naxi cultural activities and symbols have been preserved, including nightly dance in Sifang Square, Naxi ancient music performance, and Dongba pictograph exhibitions. One main reason for cultural preservation is to attract tourists. In the new city where Naxi people live, these cultural activities cannot be found in public. On the expressway of modernization, the new city emulates coastal cities to become a jungle of concrete buildings. During some local Naxi festivals, such as releasing river lanterns in mid-July, each neighborhood committee (juweihui, 居委会) in the old town organizes former native residents, now living in the new city, to return to the town and practice these festivals. As Haitun explains,

> Former native residents rarely enter the old town. Even if they do, they are more like tourists than residents, to do things such as celebrating festivals as organized by residents' committees. I think it is inevitable. Additionally, Han people are really good at business.

These former residents maintain distance from the old town, as their social and cultural relations with the town remain tenuous and exist only in memory. When the new city replicates other Chinese cities and loses its cultural connection with Naxi tradition, the old town consolidates its cultural uniqueness, not necessarily as a replication of "authentic" Naxi culture fixed in time and space, to articulate a hybrid modality that synthesizes modern and traditional, local and global. Owning a property in the old town cannot guarantee them a sense of home. Meanwhile, gated communities and high-rise apartments in the new city foster a "modern" lifestyle, in contrast with traditional neighborhoods and close-knit communities in the old town as in the past, which inevitably repeats the loss of home that tourists and lifestyle migrants endure in other Chinese cities.

In October 2019, Haitun planned to rent the shop to others after Yongxin started to run his own business. It seemed that the shop was a part of her life and embodied personal feelings, to such an extent that she hesitated to decide. Meanwhile, Yongxin was very busy and had no time to take care of their two young children. Eventually, Haitun rented out the shop in December 2019 and shifted her main attention to childcare. Nevertheless, she kept a small drinking stall on the front corner of her shop. Explaining why she insisted on maintaining this stall, Haitun notes:

> I told my business partner that I will reopen the shop one day, although I don't know when this will happen. I am not willing to stay at home to take care of my husband and children. Being a full-time housewife is not my interest. I don't want to do this. That is why I keep this stall so that I have some fun for myself.

Every day, Haitun drops by the courtyard house after sending her daughter to a local primary school and asks her mom to look after her son. Then she opens the beverage stall for business and enjoys temporal autonomy. Obviously, she contends that her business is not for money, but for fun, since she takes the opportunity to interact with other people and get some fresh information about the outside world. Her time is divided between childcare and her business hobby: "after sending my daughter to school, I finish domestic work and then open my drinking stall until her class is over at 3:00 p.m. I pick her up, send her to after-school training classes, and ensure she finishes her homework at night."

Childcare obligations pressed Haitun to close her café business, but her insistence on a beverage stall epitomizes her refusal to define her identity as a housewife. In China, females still assume the main responsibility of childcare and domestic chores, no matter whether they have their career or run business. This responsibility is particularly explicit in Lijiang. In traditional Naxi society in Lijiang, women worked arduously to support their family, while men spent more time in arts activities such as chess, calligraphy, and painting in order to emulate their Han counterparts in proper China. For instance, Naxi women traditionally wore sheepskin capes. Now the cape has given way to "a lighter, synthetic-looking version" similar in style to the original attire (Rees 2000, p. 151). Seven circular ornaments form a distinctive pattern on the capes, with the appearance of the sun, moon, and stars. This decoration is explained to symbolize female diligence, since women got up before sunrise to work and finished a busy day after moonrise. Today, Naxi men do not learn arts anymore, but play mahjong, drink, and have fun, while their wives continue to take responsibility for arduous domestic duty (Chao 2012; Su and Teo 2009). As a native Naxi, Haitun certainly understands this tradition and shows commitment to the established family value of men staying outside and women inside. Nevertheless, Haitun receives much information from tourists and migrant entrepreneurs about gender equity and refuses the label of housewife. The drinking stall becomes a tenuous bridge between her and the external world, creating a space of autonomy to mediate the gender role prevalent in Naxi society.

In October 2020, Haitun's parents still lived in their ancestral house and remained among the very few native residents in the town. Haitun's father actively participated in various activities to promote Naxi culture, while her mother unwillingly accepted interview requests. Haitun and her parents do not have a clear idea of how to handle the courtyard house. As Haitun and Yongxin told us, the two elderly parents express strong unwillingness to move out to the new city, and face no economic pressure to rent the courtyard house out to migrants. Haitun and her two children drop by the courtyard house in the old town and have dinner with the two elders every day. But they spend their nights in their own house in the new city. Nowadays, Yongxin rarely joins the family dinners or visits the old town, simply because he has been too busy to experience this sort of homely life. The temporal disjunction between Yongxin and Haitun is interesting. When Yongxin slowed down for recuperation, Haitun ran the café in full swing; when Yongxin scrambled for new business beyond Lijiang, Haitun preferred to enjoy loafing in the courtyard house. At this disjunction lies their different understandings of homely life. As Haitun reflects on the divergence,

> There is no need to coordinate our pace of life. Yongxin has a busy life and I walk very slow. How to coordinate? I cannot ask him to wait for me. Nor do I speed up to catch him. Both strategies are unfeasible. We need to understand each other. He wants a busy lifestyle to fulfil his dream, and I understand and support his choice. I prefer to sit in the courtyard to bask in the warm sunshine, and he understands that too.

Home for Haitun is embodied in a representation of cultural values through the town, the courtyard house, and the natural environment, and an experience of spatial and temporal autonomy by her drinking business, slow pace, and sunbathing. While Yongxin gravitates towards the frivolous and speedy journey for wealth, eager to become a boss of his own to test his entrepreneurial skills after seven years of recuperation and rest, Haitun makes efforts to maintain her own art of living in the old town and keep distance from busyness, although she takes the main responsibility of childcare. If Yongxin wants to project his entrepreneurial ambition into the world and embrace the race for wealth and fame, Haitun prefers to enjoy a reclusive style. Together, the couple fosters "one home, two livings."

Furthermore, the mutual understanding emphasized by Haitun in relation to Yongxin does not necessarily indicate an overt resistance to patriarchal order in China, but highlights that she can be herself at home. This feeling of being at home forever, as Haitun arduously maintains in her life, enables her to withstand the loss of home that Yongxin and many others in the old town try to escape. Due to this feeling, Haitun does not suffer from in-situ displacement that other native residents must cope with. Surrounded by unhomely subjects coming into and departing from the old town, Haitun rekindles the value of home in a relentless world through a delicate balance between domestic responsibility, spatial-temporal autonomy, and low materialistic desire. By detaching from the race for wealth, Haitun practices the art of life inherent in traditional China. Indeed, Haitun's stay in Lijiang Old Town cultivates an experience of reclusion. Regarding the lifestyle choice, Lin (1998, p. 160) specifies that "the great Taoist teaching is the emphasis on being over doing, character over achievement, and calm over action. But inner calm is possible only when man [sic] is not disturbed by the vicissitudes of fortune." Haitun's life journey explicitly shows that she has sufficient mastery over her mind and feels unafraid of her surroundings. She is a great recluse even in a bustling place like Lijiang.

Conclusion

This analysis of stay and displacement by native residents of Lijiang Old Town highlights the need to study the continuity and disruption in China's rapidly changing society. The large flows of tourists along every street in Lijiang Old Town brought noise, crowdedness, and gaze to disrupt and constrain native residents' daily routines. Meanwhile, rising property rents pushed them to calculate the pros and cons of continuing to live in the town. Different from lower-class residents in those gentrifying neighborhoods who suffer from ubiquitous harassment and even violent intimidation, these native residents had adequate agency to decide what to do once the old town, where they grew up and were familiar with, turned into a strange land filled with tourists and migrants. Technically, no one could evict them from their own property, but the town has become a strange and uninhabitable place for them.

A focus on the uncanny conditions of native residents feeling strange in their hometown illustrates the synthesis of the race for profit and the desire for home to shape the space of daily life. Despite the fact that they can control their property and face no pressure of eviction, native residents have been struggling to maintain their claims to the town: who can enjoy the town as a home and who cannot.

It seems that a trade of meaning for money is widely embraced by young Naxi people who eagerly catch up with China's first-tier cities such as Beijing and Shanghai, while the elderly have made great efforts to prioritize meaning over profit as they grapple with the deal of modernity. As Young (1997, p. 164) reminds us, the home is imbued with "a nostalgic longing for an impossible security and comfort, a longing bought at the expense of women and of those constructed as Others, strangers, not-home, in order to secure this fantasy of a unified identity." This longing is evident to native residents in the surge of disparate forms of connection and separateness – from the evasion of harassment to the dream of normality to the quest for true love – for the goal of a "return to familiarity and order of everyday life" (Rojek 1997, p. 70). Once this return cannot be fulfilled, native residents in the old town choose to leave for the new city and uproot themselves. This analysis of how their rooted home is lost in Lijiang Old Town generates some insights into the dynamic affinity between the subjective idea of self-development and the objective world of rational efficiency. Indeed, the pursuit of home by tourists and migrants relies on the loss of home by native residents to turn the old town into an unhomely space in which the former escape modern cities to enjoy a homely life in Lijiang through their purchasing power, while the latter give up their homes to resettle in a modernizing new city.

In *Modernity and Ambivalence*, Bauman (1991, p. 94) expands the discourse of the stranger to emphasize the universality of rootlessness: "[T]he mode of 'being a stranger' is experienced, to a varying degree, by all and every member of contemporary society with its extreme division of labour and separation of functionally separated spheres." The result is that individuals are socially displaced and uprooted from a place and do not feel at home. This displacement happens to native town residents and eventually, an overwhelming majority of them choose to leave the old town and move to the new city, a symbol of modernization and urbanization in China. My analysis of what happens to native residents in Lijiang, along with their struggle to feel at home in the town, is part of a call for attention to the social and mental desire of the elderly who hold their home firmly as a solid entity to mediate a rapidly changing society. It also generates a much more expansive conversation about how the race for wealth pushes those native residents into an unhomely condition, even though they stay away from the race. This analysis does not repeat an understanding of gentrification as a class-imposed accumulation by dispossession (Harvey 1982), but highlights a prevalent form of trade between profit and meaning. While much has been written about the injustices of capital-driven urban redevelopment to accelerate gentrification and forced eviction, what happens to native residents in Lijiang Old Town offers a nuanced picture of urban transformation that is driven not merely by capital accumulation, but also by place attachment, age and gender difference, and social bond.

It is easy to spotlight the pervasive and indomitable aspect of fluid modernity and therefore exaggerate the tendency toward transiency and uprootedness (Fog Olwig 2002), but the insistence on staying in Lijiang Old Town by a few residents demonstrates their strong orientation toward their place of origin, although it has become a site of tourism production and consumption. The four stories of making and remaking home shed light on struggle and adjustment at the micro level. The locals who choose their best ways of homemaking are not only responding to external challenges and opportunities brought by Lijiang's tourism development, but also advancing their own vision of the deal of modernity. The figures in these stories delicately handle the balance between money and meaning so that they can adjust their position and remain resilient in the raging torrent of China's socioeconomic transformation. It is easy to categorize their adjustment efforts as the struggle for a better future or the hope of a just society. Nevertheless, this categorization cannot fully comprehend their choice of living with mobilities or the internal struggles for inner peace and cultural roots in relation to Lijiang's dramatic changes. Rather, these stories exemplify different forms of human agency in response to the process of modernization. Regardless of their choices of departure or stay, native residents in Lijiang Old Town cannot be read simply as passive recipients of China's great transformation, nor victims of massive commodification driven by Han migrant entrepreneurs. This reading is one-sided and can distort the complicated picture of what happens in Lijiang. Rather, their departure or stay represents a rational response to the deal of modernity, through which they attempt to strike a balance between money and meaning. In the next and final empirical chapter, I turn to local service workers to explore different struggles and adjustments unfolding in the deal of modernity.

Note

1 This interviewed was conducted in 2004. In 2019, the rent could have jumped to CNY 150,000–250,000 per year.

Chapter Six
Hometown Babies: Immobility and Lijiang Locals' Struggles for Home

Ever since I started to conduct fieldwork in Yunnan province in 2002, I have heard many times the term *jiaxiangbao* (literally, hometown babies, 家乡宝) to describe Yunnanese people's reluctance to leave their hometown. Many of my respondents in Lijiang, Kunming, and more recently, Banna and Ruili, laughed at their predilection for staying in Yunnan and even in their birth town, and admitted that they lacked the strong ambition of those in big coastal cities. To them, *jiaxiangbao* implies two things. First, they love their hometown and enjoy themselves in Yunnan. Similar to babies attached their mothers, they stick to their hometown and develop a strong sense of attachment to it. Second, this strong sense of attachment discourages them from looking outward for new opportunities and achieving greater success, particularly in China's economically vibrant coastal cities. Historically, Yunnanese people, mostly driven by poverty and political upheavals, arduously established business networks in mainland Southeast Asia to fulfill their entrepreneurial ambition. Chang's (2014) work on Yunnanese migrants in Myanmar highlights their experiences and transnational trading and risk taking. Today, some Yunnanese continue to forge ahead all over the world, but the majority choose to become *jiaxiangbao*. In his analysis of returned migrants in Kunming, for instance, Werner (2012) uses *jiaxiangbao* to describe Kunming locals who choose to return to their hometown even after they have studied abroad and worked in Beijing or Shanghai. Three major factors identified by Werner (2012) include comfortable weather, filial piety, and intense social networks formed in childhood. These factors are similarly cited by Lijiang

Unhomely Life: Modernity, Mobilities, and the Making of Home in China, First Edition. Xiaobo Su.

locals when they explain why they prefer to stay in the valley instead of migrating to other Chinese cities outside Yunnan, although this stay is fraught with struggles.

Hometown babies are akin to what Heller (1995) calls geographically monogamous persons. This form of geographic monogamy represents a spatial formation of *xiangtu* society in China. As Fei (1946, pp. 1–2) points out, "peasantry, the key toward understanding China, is a way of living, a complex of formal organization, individual behavior, and social attitudes, closely knit together for the purpose of husbanding land with simple tools and human labor. Peasants are settled and sedentary." Thus, everyday life was not conceived of as separate from leisure and work, as people were bound to the soil and fixed in one locality. Homemaking remained a relatively stable totality of human practices in a familiar setting. In this case, we can argue that rural villagers in ancient China could be regarded as hometown babies, because they had been bound from birth to death in their home villages. In Lijiang, hometown babies identity their hometown as the central point of their life, although the definition of hometown ranges from a village to a prefecture and to Yunnan province. This is particularly true in Lijiang since locals rarely migrate far to other cities for a living. They prefer to live their whole life in their own hometown, partly because of a very strong sense of place attachment and partly because they get used to their own lifestyle. If they must leave Lijiang, the only place that they can think about is Kunming, the capital city of Yunnan. Any city outside Yunnan is hardly imaginable. Hometown babies signal a form of immobility, and in an era of great transformation, they illustrate both value and challenge as far as homemaking is concerned.

Here I define urban Lijiang as the core built area – the new city and the old town, while rural Lijiang refers to towns and villages where agricultural production plays a dominant role in social and economic development. Since the development of tourism in the late 1990s, increasing numbers of locals have chosen to relocate from rural Lijiang to urban Lijiang. Many of them participate in the tourism sector by offering necessary services such as cleaning, express delivery, and restaurant service. "Locals" refer to those who live in both rural and urban Lijiang, and self-identify as local people (bendi ren, 本地人).[1] Many locals in urban Lijiang do not make a living from the soil or engage in farming, since they rely on the tourism industry to earn their livelihood. Nevertheless, rurality and traditional culture continue to shape their sense of time and space, and closeness to nature is the root of their life.

The difference between locals who come from distant rural villages and those who dwell close to the old town is explicit. To Lijiang locals who originate from distant rural villages (two or three hours by bus, or even farther) and "migrate" to live in urban Lijiang, there is no need for them to rely on husbanding the land for agricultural production, since they are integrated into the local tourism industry and make a living from their service. Nevertheless, the cultural sentiment born out of peasantry continues to shape their way of life and pushes them to settle down. Perhaps their next generation can become fully urbanized and sever their ties with the land, but their own life journey starts from the land, which has been unattended due to low economic productivity, and the settled and sedentary way of life is deeply imprinted in their daily practices. Literally, it is difficult for them to return to husbanding the land, although people scattered in the valley gather on ceremonial occasions. But culturally, they persist in a rural way of life and struggle to become modernized in urban Lijiang. The separation between economic engagement in urban Lijiang and cultural sentiment in rural Lijiang implicates these locals into an unhomely condition.

Those who live close to the old town endeavor to maintain their established pattern of time and space. They build spacious courtyard houses on their own homestead sites and choose to earn a living from the tourism industry. They stay in their own house and have autonomy to manage their time. Undoubtedly, this pattern is changing due to tourism development and other projects of modernization in Lijiang. If the old town has dramatically evolved into a Disneyland-like playground for tourists and lifestyle migrants to experience a homely way of life, the local society outside the town develops its own pace of transformation. The new order of modernity is mediated by established patterns, resulting in a gradual change of the habitual way of living by Lijiang locals.

Speed and Slowness: The Supply of Homely Service to the Old Town

Lijiang's tourism industry relies on service workers who mainly come from the valley. Here I focus on three main groups who display different patterns of time-space: female cleaners who work in guesthouses, male workers collecting and delivering parcels in the town, and finally, those who offer rides to tourists. Generally, the first two groups are subject to time discipline, while the final one has more temporal autonomy.

Guesthouse A'Jie and the Commodification of Domestic Work

Lijiang guesthouse owners often rely on one manager (guanjia, 管家) and one service worker (A'Jie, literally, senior sister, broadly referring to married women in Lijiang). Of the forty guesthouses I have investigated since 2015, only one does not hire a manager due of its small number of bedrooms. Most managers come from other Chinese cities, are well educated, and understand tourist-related business. Half of them are male, and many begin their career as volunteers in guesthouses. Their main responsibility is to handle online reservations, welcome tourists, organize tea drinking, supervise A'Jie to maintain the guesthouse, and so on, with a primary purpose to serve tourists directly. As guesthouse owners become busy (again) in exploring new business opportunities (Chapter 3), managers take responsibility for creating a homely environment for tourists. Managers and owners develop rapport under the same roof, since they have meals together every day, share business information, and even travel together. Some managers also are involved in owners' other business such as online sales and livestreaming commerce.

In the past, guesthouse operators hired xiaomei (literally, young sister, or teenage girls, 小妹) from villages in Lijiang's vicinity. These xiaomei got a monthly wage and conducted various manual tasks such as housekeeping, cooking, and attending to customers. Since the 2008 financial crisis, xiaomei either went to Kunming for better jobs or pursued college degrees, which means that they became less available. Gradually, guesthouse owners started to hire middle-aged women, or A'Jie (阿姐), to conduct maintenance work. Today, xiaomei exist only in bars in the old town, where a higher salary is offered, and guesthouses rely on A'Jie. Most A'Jie come from rural Lijiang and some from urban villages in Lijiang's new city. Generally, a guesthouse A'Jie is responsible for preparing one or two meals (lunch and/or dinner), watering flowers, and cleaning the guesthouse – a combination of chef, janitor, and gardener.

There are two types of guesthouse A'Jie. Nonresidential A'Jie live outside the old town in either their own house or rented rooms, and commute daily to work between 8:30 a.m. to 4:00 p.m. Residential A'Jie live in the guesthouse and have meals with other staff members. Their working hours start around 8:30 a.m. to shop in local wet markets, and end around 6:30 p.m. after dinner. They are occupied with maintaining the guesthouse, but generally take a break before dinner preparation if maintenance is completed. After dinner, residential A'Jie spend their time shopping, meeting friends, or relaxing, and go to bed around 11:00 p.m. During their off-duty hours, residential A'Jie avoid the main area where tourist customers lounge. They either go out or stay in their bedroom. A'Jie can take a one-day vacation per month. No extra benefits exit to compensate their work on weekends. They can ask for a unpaid leave to handle personal affairs, but cannot be absent frequently.

In their late thirties to fifties, A'Jie provide essential support to guesthouses and maintain a homely environment for both owners and tourists. Their job encompasses a range of temporal and spatial uncertainty that extends beyond a fixed pattern of work hours. During the high season of tourist arrivals in the town, they rush to clean bedrooms for tourists seeking cozy and comfortable space. During low seasons they seek other temporary jobs for cash. Generally, housekeeping A'Jie can earn monthly between CNY 2500 (~US$ 390) and CNY 3000 (~US$ 465), depending on how long they work for various guesthouses. With the agreement of the guesthouse owner, some A'Jie can work as a part-time janitor for other guesthouses, earning CNY 15 (~US$ 2.21) per bedroom. Generally, housekeeping A'Jie provide much-needed service to facilitate the loafing lifestyles enjoyed by tourists and migrants, and replicate their own domestic work in other people's abodes, giving rise to the commodification of domestic service.

A'Jie come from either rural Lijiang or the town's surrounding areas. For instance, A'Jie Helei is of Naxi ethnicity and is around forty-five years old. Her marital village is about one hour from Lijiang Old Town by public bus. Her son is a guesthouse manager in the old town, and her husband is a self-employed driver providing chartered car service for tourists. Her daughter worked as a bar attendant in the old town before getting married in early 2020. A'Jie Helei and her husband rent a bedroom in an urban village two miles away from the old town. Her whole family lives in urban Lijiang and makes a living from the tourism industry. In another case, A'Jie Chen, in her forties, lives in her own house in an urban village several miles away from the old town. Her work hours are 9:00 a.m. to 3:00 p.m. In the mid-1990s, she started as a construction worker in the new city. After a serious injury, she could not bear any tough manual work; since 2006, she has begun to work in guesthouses as a nonresidential A'Jie. After working and saving very hard, A'Jie Chen and her husband were able to build a large courtyard house on their own homestead, with eleven bedrooms in total. Due to the short distance to Lijiang's downtown, they can rent some bedrooms to outsiders who work in urban Lijiang. In October 2020, she rented out two bedrooms, CNY 500 (~US$ 73.75) per room per month, so that she has some extra money.

Albeit at the bottom of the labor market, A'Jie have their own dignity and strive for mutual respect with tourists and guesthouse owners. In one guesthouse where A'Jie Chen worked previously, the owner often asked her to do extra work beyond the agreement. For instance, the owner liked drinking and sometimes got drunk and messed up his own bedroom. Then he asked A'Jie Chen to clean the mess, which made her very unhappy. Normally, A'Jie has no responsibility to clean the guesthouse owner's bedroom because this task can make them feel treated as servants instead of cleaners. Cleaning up after the owners can happen once or twice as a token of reciprocity, but A'Jie Chen realized that the owner did not respect her. Later, she quit the job and found her current one. She comments,

> Mutual respect is important to both the boss and A'Jie. The boss cannot insult my dignity or treat me like a servant. The current boss fully respects me and trusts me very much. I feel very good. For instance, he invites me to join the tea table to drink tea together with tourists, which rarely happens in the town. When he and his wife travel to Tibet for ten days, they leave the guesthouse to me for operation. I also invite the couple to my own courtyard house for barbecue. Sometimes I bring vegetables from my own garden to the guesthouse and share with the boss. We get along well with each other, like friends under the same roof.

A'Jie Chen seems quite content with her living conditions. As she emphasizes, she works as an A'Jie not merely for money. Her two daughters have graduated and will launch some new business in the near future. Her husband works as a construction worker with good wages, due to Lijiang's tourism-driven real estate market. Being an A'Jie, she can earn pocket money and, more importantly, have some things to do and encounter various tourists from other parts of China. Interaction with tourists can enlarge her horizon and help her to receive new information about the external world. After work, she either meets her friends in the new city or goes home to take care of her garden. She has no intention of doing extra maintenance work for another guesthouse. Neither does she want to extend her work hours for more money. A workday from 9:00 a.m. to 3:00 p.m. allows her to finish assigned tasks, and gives her adequate time for personal activities.

> The more money you have, the more you can spend. If you earn less, then spend less. We need to maintain good health, don't we? In the past, I worked very hard to earn money. The hardship my husband and I endured was beyond imagination. Now it is time for me to enjoy a little bit. Life is short and those in their forties start to suffer from various diseases. Some fellow villagers who were younger than me suddenly passed away. I am really shocked. What is the meaning of life if you work too hard, but have no time to enjoy yourself? In pursuit of wealth, there is no ending point. Enough is enough.

The work–life balance that A'Jie Chen describes enables her to stay away from the rat race for wealth and enjoy a homely life in her own hometown. Having said this, her choice depends on the fact that she and her husband can earn adequate salary for a living and face no pressure to buy a commercial apartment in the new city.

While A'Jie Chen is quite exceptional in terms of the deal of modernity, most A'Jie I encountered in Lijiang prioritize monetary earning or save money for the future. Thus, they work very hard, swiftly rotate from one task to another, and live a frugal life. Some even rush from one guesthouse to another to clean as many rooms as possible, particularly during Lijiang's high tourism season (from May to November before the COVID-19 pandemic). Bedroom maintenance and meal preparation keeps them busy. This contrast between slow tourists and speedy workers is akin to what McClintock (1995) finds: in Victorian times the British cult of domesticity for an ideal home was underpinned by the demand for domestic labor. McClintock (1995) uses a "doubling" of women to describe an odd encounter between the idleness of middle-class women and the degradation of working-class women. This encounter also happens in Lijiang, albeit in a different manner. While household chores in guesthouses are outsourced to female service workers, tourists and operators can enjoy loafing in the town and experience a homely lifestyle. The gendering of consumption and production becomes part of homemaking in Lijiang.

Delivery A'Ge and Time Discipline

In addition to A'Jie, another important group of service workers in and around the old town are middle-aged men, called A'Ge (阿哥), whose main job includes express delivery in the old town. Three delivery firms with offices in the old town hire about twenty delivery A'Ge, most of whom come from rural Lijiang. During work hours, delivery A'Ge are extremely busy in collecting and sending parcels between customers and their operation office. Due to strict regulations on heritage preservation, these delivery A'Ge are not allowed to use any machines (bicycles, tricycles, or carts), but must conduct their work manually. If they are caught using carts or tricycles, the penalty is about from CNY 5,000 (~US$ 737.46) to CNY 6,000 (~US$ 885). Thus, they use backpacks to carry parcels and walk between the operation center and customers, which makes their work exhausting and hard. They need to arrive each day at the distribution center in town at 8:00 a.m., and their main duty includes conducting three town-wide trips to deliver and collect parcels. Work ends at 10:00 p.m. Generally, each delivery worker walks around thirty-thousand steps per workday, the equivalent of about 24 km.

Underpinning the old town's homely lifestyles are delivery workers who rush to facilitate a consumerist society. A'Ge Huang is a middle-aged man who collects and delivers parcels in the old town. After taking his children to school, A'Ge Huang goes directly to work. He must pay a monthly mortgage and supply his two children's tuition and pocket money. Due to the rise of livestreaming sales in the old town, the express delivery business is booming. Nevertheless, the job is too exhausting to attract young workers. As A'Ge Huang explains, "Young men do not take this position in the old town's distribution center. It is too exhausting and overloaded. There is only one day off per week. If I were young, I wouldn't do this job either. Because I must support my family, I have no choice. Middle-aged men have no choice." The sentiment is shared by other delivery A'Ge who dread the hectic pace of their job. After ten minutes of conversation, for instance, another delivery A'Ge, Siyu, says that he cannot talk any more otherwise his manager would be mad that he was wasting time. Then he left in a hurry to deliver parcels to customers in the old town.

Despite the hardship, delivery work can generate a relatively good salary compared to other service jobs in the old town. Delivery workers' income includes two parts: a base salary and a package deduction. The base salary is very low, and delivery workers deduct CNY 1.8–2 (~US$ 0.27–0.29) from every package they either collect or deliver. The more they deliver or collect, the higher the deduction they earn. Time is money, and efficiency is life. Their time is subject to the logic of time discipline, although they can choose to slow down in the town. On average, a delivery A'Ge can earn CNY 8,000–10,000 (~US$ 1,180–1,475) per month, much higher than for a guesthouse A'Jie. Work pressure pushes them to walk fast and work long hours; relatively high earnings come at a price of physical hardship and heavy overtime.

Delivery A'Ge are more engaged in time discipline than any other groups of respondents I studied in Lijiang. Similar to their counterparts in China's coastal cities, they submit to the ebb and flow of time in a town known for loafing and comfort. Unlike tourists who consciously choose to slow down and stroll in alleys without purpose, delivery workers must speed up and develop an accurate map of the town so that they can serve their customers without wasting time. No one is more familiar with the town than these delivery workers, yet they are alienated from it more than any other groups of respondents. They offer great economic and cultural value in Lijiang's home-based experience economy, with a hope for a better home through their self-disciplined use of time and space for more economic return. Since their labor force can be easily replenished, their employers, headquartered in Shanghai or Shenzhen, have no reason to endow delivery workers' time-space with meaning. Neither does Lijiang Municipal Government care about their physical and mental wellbeing, since new migrants such as Yongxin can bring more economic value to local development.

These delivery workers are connected to a command center unit via cell phone apps. Their routes and paths are mostly located around the old town, hence their familiarity with the labyrinth of alleys and streets. The rate-per-piece system suggests that the race for more deliveries pushes them to work harder, and faster. They must deliver goods to their customers in time and at the right spot, otherwise they receive negative comments and lose much-needed credit. They are disciplined for tardiness, or for mistaking a customer's location. Security guards employed by Lijiang Heritage Authority yell at them and treat them like second-class citizens for disrupting the town's loafing atmosphere. They are subject to stern evaluation by the employer, verbal violence by security guards, and ignorance and contempt from their customers. In the midst of all of this, they are uprooted from the town and suffer a loss of home in a place they are so familiar with. To them, a sense of familiarity is not akin to a sense of home. Perhaps they are too familiar with the town to feel at home in it.

In *Discipline and Punish*, Foucault (1979) notes that the use of time is incorporated into a disciplinary arrangement upon the body to create conditions of efficiency and speed. As Foucault (1979, p. 152) argues, "In the correct use of the body, which makes possible a correct use of time, nothing must remain idle or useless; everything must be called upon to form the support of the act required." These delivery workers' time-space in the town is dictated by algorithmic scheduling and fragmented by consumer demand, giving rise to the subsumption of labor to machines and profit. Indeed, these workers trade meaning for money, once they relegate the control of their own time and space to abstracted webmasters in charge of delivery orders. Always rushing to deliver for their customers shapes their sense of time and space. They spend most of their time in the town, but the time has no specific meaning. If there is a meaning, it is about how to credit out for actual economic return. In this sense, no dramatic difference exists between these workers and those in China's hustling cities in terms of the logic of time discipline. While the latter can travel to Lijiang Old Town to experience a homely life, the former must rush in the town to support such an experience for others. The structures of capital drive delivery workers into the old town to facilitate the home-based experience economy, first displacing them from their home villages and then displacing them from the town. Biopower is exercised upon delivery workers through time discipline and spatial displacement.

Freelance Workers for Tourists

The third group consists of drivers who offer necessary transportation service to tourists. The service includes shuttle rides between the old town and the airport or railway station, online car-hailing service, and tricycle rides between the town's entrances and guesthouses. These drivers can be female or male, and most of them come from local villages close to the old town. Most drivers are officially registered to serve tourists, and some run their business by oral agreement.

A'Jie Wangli is a female driver of Naxi ethnicity. She and her family live in a village close to the airport, 30 km away from the old town. As a driver affiliated with Didi, China's biggest online car-hailing company, she mainly serves tourists. This job gives her much temporal autonomy, since she can decide when to go online and activate her service. Thus, she can earn money from the tourism industry and take care of her family. Early morning, she sends her children to school and starts to work. Around 4:30 p.m., she stops driving, picks up her children, and goes home. Thinking about how tourism development shapes her life, A'Jie Wangli notes,

> Thanks to tourism development driven by the old town, we also benefit from it. When people come to visit the town, we have new business opportunities and make money. There is no need for us to do business inside the town. Rather, local people can operate hotels or restaurants [outside the town] to cater to tourists. Some can work for online car-hailing companies such as Didi.

From this quote, one can discern the positive economic benefit brought by tourism development to local communities in Lijiang. In the 1990s, very few Naxi females worked hard driving taxis to earn money (Chao 2012), but today, increasing numbers of local females participate in the tourism industry by driving cars to serve tourists. Through her freelance work, Wangli can take care of her children and thus fulfill her duty as a mother.

In a different case, A'Ge Gao is a driver of Bai ethnicity and affiliated with a tour company. He is married and still stays with his parents in a big courtyard house in a village 10 km east of the old town. His daily schedule revolves around tourism. If he is tasked to serve tourists, then he offers a ride. If there is no task, he can spend his time at will. He enjoys "hanging with friends and relatives" for fishing and barbecuing. Despite tourism development, A'Ge Gao and his family continue to enjoy a lifestyle they prefer, including slowness, relaxation, barbecuing, and a spacious courtyard house. This lifestyle contrasts with the one that tourists experience in big cities. A'Ge Gao notices this difference and sometimes feels baffled at the pursuit of speed by his tourist customers. He tells one story:

> Once I gave ride to a tourist from Shenzhen. He asked me whether I can drive a little faster. I was curious and asked back whether he was in a hurry. He said no. I further

> asked why we need to ride faster? He replied that he was used to fast paces in Shenzhen and felt quite uncomfortable in a low-speed car. Then he murmured that our living pace [in Lijiang] is too slow. It is true, and I love it.

Gao's observation illustrates the long-neglected reality that speed has become internalized in China's big cities such as Shenzhen, and people spontaneously get used to the cult of speed. The internalization of linear, abstract time, as Rosa (2013) asserts, transforms how time is used and experienced in the domains of work and leisure. The contrast between Shenzhen and Lijiang is reified through a different need to accelerate or decelerate, a contrast turning Lijiang into a site of mediation between systemic imperatives and subjective perceptions by way of a social construction of time (Rosa 2013).

While Chapter 4 elaborates how retreat tourists enjoy loafing in Lijiang Old Town and slow down their pace, many other tourists still indulge in speed, by rushing to eat, to visit attractions, and to take pictures, unable to break free of the manacle of time discipline. Undoubtedly, A'Ge Gao's slow ride caused a stir to the tourist from Shenzhen. In Lijiang, Gao prefers to follow the established pace of living and, to some extent, pushes his customer to enjoy this pace. Ironically, Lijiang locals' slow pace has been widely regarded by tourists and even lifestyle migrants as a form of laziness and unproductivity.

In addition to car drivers, tricycle riders play a role in helping tourists and business owners in the old town. Through a business relationship, these tricycle riders either deliver goods to restaurants or cart luggage for tourists (Figure 6.1). For restaurants, they are tasked to deliver goods such as food ingredients or drinks regularly. For guesthouses, they will receive a call from the owner to cart heavy luggage for tourist customers. Or they simply wait in the parking lot at the town's entrances to wait for tourists. The cost per ride is CNY 20–25 (~US$ 2.95–3.69), depending on the distance. In the high tourism season, these riders can do five to eight trips and earn CNY 100–160 (~US$ 14.75–23.6) per day. In the low tourism season, however, their earning is much less. The old town has five main entrances, and each entrance stations around 10 tricycle riders. These workers are registered with Lijiang Heritage Authority so that they can ride tricycles for business in the old town.

In his fifties, A'Ge Heli comes from Yuhu village, 20 km away from the old town. To stay longer for more rides, A'Ge Heli rents a bedroom in an urban village in the new city. Heli's wife works as a hotel janitor in the new city, and her working hours are routinized. She commutes everyday between Yuhu village and the hotel so that she can raise chickens and pigs and maintain their unfurnished house. Previously, Heli was a security guard in the old town. Then he quit the job partly because the salary was low and partly because he had to work every day. After becoming a tricycle rider to serve tourists, he can earn more and have a flexible schedule: starting to wait in the designated entrance at around 10:00 a.m. and going to sleep at 10:00 p.m. if there is no business. He

Figure 6.1 A tricycle rider in Lijiang Old Town in 2018. **Source:** Xiaobo Su (Book Author).

is quite satisfied with his earnings, but worried about the seasonality of tourism. Tricycle riders like Heli can be regarded as low-end freelance. Theoretically, they can manage their time at will and enjoy temporal autonomy. Having said this, their time-space is subject to the demand in the market, and their earning is associated with how much time they spend waiting for customers and riding tricycles.

Free Time, Away from Lijiang Old Town

While Lijiang Old Town has been regarded and even idealized as an oasis of retreat and recovery by lifestyle migrants (Chapter 4) and tourists (Chapter 5), Lijiang locals who engage in service work have a completely different opinion. When Lijiang locals take a rest from busy work, they utterly refuse to visit the town, but rather spend free time in the new city for relaxation. As A'Ge Song explains, "If it is not for business purposes, we never want to walk into the old town. During the days of rest, no one comes here. I am so familiar with the town, and pretty much know every flagstone on streets and alleys." Song's explanation indicates that the town does not represent a space of relaxation and loafing, as cherished by tourists, but rather the chains of time discipline upon their labor. Asked whether he wanders in the old town, A'Ge Gao said: "No, absolutely no. It doesn't make sense to me. People visit it to buy some souvenirs. The town might be unique to tourists, but to me, it is too ordinary. If it is not for my job, I never want to enter it."

Like many other local service workers, A'Jie Wangli considers it ridiculous to stroll in the old town like tourists. She and her friends firmly decline to enter the old town. They perceive a clear boundary between the old town and the new city, since the former is reserved for migrants to run business and for tourists to have fun, while the latter remains a modern site for activities akin to those in China's big cities. As Wangli explains, "it is meaningless to stroll in the town. Nowadays we love singing. If we have time, we go to KTV or gather for meals outside the town." Although the old town is a pillar in Lijiang's tourism development from which Wangli and other locals earn their livelihood, the boundary puts them out of place. In the old town, these locals become outsiders.

This distance has two implications. First, local service workers want to control their own time-space once they are off duty. Loafing remains one of the favorite ways to experience time in China. In the old town, tourists and migrants exercise their purchasing power to buy the moments of loafing over service workers' space and time. I emphasize that this experience by relatively affluent lifestyle migrants and middle-class tourists passes the burden of time discipline to those services workers who rush to finish one task after another in order to keep the old town running smoothly. While the former group attempts to evade the 9-9-6 race at home and look for ways to waste time through a homely way of life in Lijiang, the latter cannot withstand the encroachment of speed upon their time in the old town. By staying away from the old town when they are off duty, these service workers manage to add meanings to their "new home" in Lijiang and care about their family members.

Second, they rarely regard Lijiang Old Town as their home, even though they are quite familiar with it. If we shift gears and look inside the experience economy, then we see that under the surface of loafing and comfort that tourists and migrants enjoy is a parallel world of time discipline upon service workers who involuntarily choose to work hard and fast for wages. Due to the dictates of algorithms and machines, they feel so alienated that they refuse to stroll in the town when they are off duty. For them, the town is a space of speed where they are integrated into the production of experiences, instead of a space of relaxation where they spend time for nothing. Except for some necessary interactions with tourists, they remain invisible and detached from the old town. The distance indicates a social and spatial boundary between Lijiang locals and those who come from other parts of China.

Some exceptions do exist, particularly among those residential A'Jie. Since they spend most of their time in the guesthouse, they do not have much time to step out of the old town. In her late thirties, for instance, A'Jie Li has worked as a guesthouse cleaner or restaurant attendant in Lijiang Old Town for almost fifteen years. In early 2020, she started to work in the current guesthouse as a residential A'Jie.

> I have spent most of my time in the old town, ever since I started to work over ten years ago. I rarely go to wander in the new city and am not familiar with it. Sometimes I get lost in the new city. Quite scary! But I know the old town quite well and never get lost.

Lijiang locals' service provides an interesting counterpoint to the salience of the experience economy, since their labor is precarious and dependent. Under the logic of capital accumulation, these service workers make an affective contribution to a sense of home enjoyed by retreat tourists and lifestyle migrants. The class difference between tourists/migrants and local service workers is explicit. While the former enjoy qualitative leisure in the guesthouse to mark their wealth, the homely milieu depends on the use of service workers' domestic labor of housekeeping and express delivery. An experience of homely lifestyles in guesthouses is predicated on the exploitation of service workers, invisibly present in this space to produce and maintain the comforts of home. Pratt (2004) describes domestic workers as modern-day slaves or indentured laborers, particularly in the service industry. The service workers I study in the case of Lijiang have more autonomy to carve out a relatively better condition. At least, they are not bound to any master for survival, but believe that their hard work in the town can earn a better home for them and their next generation.

What is the relationship between the old town and the rest of Lijiang? While the old town needs the rest of Lijiang to foster a cultural soil of relaxation and loafing in which the town can be constructed as a distinct and relaxed home to accommodate migrants and tourists, the new city in turn depends on the old town for job opportunities and information flows through which Lijiang locals can strike an acceptable balance between closeness to their own home and making money from the tourism market. By keeping a distance from the old town, these Lijiang locals maintain their temporal and spatial patterns and gradually absorb external influence into their daily life. They do not reject new information or ideas from China's coastal cities. Neither do they ignore the income gap between them and those on the coast. Nevertheless, their inherent cultural value of staying at home in Yunnan province makes them content with what they have already obtained so that they can reject time discipline as well as the race for wealth and fame. As they choose to keep distant from the old town, they also deliberately exclude migrants from their social networks. Eventually, the old town becomes an experiential home for tourists and migrants, while the broad area outside the old town continues to accommodate Lijiang locals. To these locals, their home is neither ossified nor idealized, but subject to social and economic relations in and beyond Lijiang.

Pain and Joy: Embracing Hometown in Lijiang

As to the question of a place called home, the difference between those locals who live close to the old town and those who come from distant villages is quite explicit. Furthermore, the struggle between stay and departure is shaped by various social factors, including gender, family structure, and age. Instead of applying one generalized struggle to my respondents, I prefer to present a complicated picture in which they figure out the relationship between the self and the world, and mobilize their agency to pursue a homely life within external constraints.

The Shadow of Patriarchal Society

As McDowell (1999, p. 78) points out, women, particularly married women, were constructed as in need of control to become the angels of the domestic arena whose duty was to bring order and warmth to their family members, showing a parallel between "women's bodies and the social construction of femininity and the home." This parallel is particularly true in Lijiang. Various social norms exist to regulate women's mobility in Lijiang's local society. Marriage seems like a turning point. Generally, an unmarried woman can stay alone outside her parents house for work in the old town, or become a rural-to-urban migrant in China's coastal cities. After getting married, local women are discouraged from migration or traveling alone. Patriarchal norms in Lijiang's traditional rural society deprive local females of the opportunity to travel afar and fix them into a local home imbued with patriarchal power.

Asked whether she wants to migrate to coastal cities for jobs, A'Jie Wangli admits that she cannot do so anymore for two reasons. First, she cannot leave her two young children for jobs elsewhere and second, it is rare for married females to become rural-to-urban migrants outside Yunnan. In her village, people show contempt toward married females who do not stay at home, but become part of a floating population. Homemaking entails a moral barrier that hinders Wangli's spatial mobility, but the tourism industry allows her to earn money locally so that she can take care of her family. Reflecting on her fate, she laments that she got married at the age of twenty one, too early to have chance to travel to coastal cities and explore a different life possibilities. Her time revolves around school hours and tourist schedules, and her daily routines swing between the new city and the rural village. A'Jie Wangli's story shows that local women's obligation to make home for their family creates barriers to their mobility. "Patriarchy" and "home" mutually consolidate each other to fix local women, particularly those married, in Lijiang. Their participation in the tourism industry enables them to evade patriarchal norms in their own community and gain some temporal and financial autonomy. Nevertheless, they are still subject to the local patriarchal order, as they cannot travel alone or stay overnight in hotels.

This sort of unwillingness is shared by other local females. Reflecting on her life journey, A'Jie Chen felt sorry for her limited traveling experience. Kunming is the most distant place she travelled to when her daughters attended schools there. During her daily work in the guesthouse, she frequently interacts with tourists from other provinces and listens to them about what happens in big cities. Unsurprisingly she hopes to visit other places:

> Locals of my generation rarely travel beyond this valley, let alone Yunnan. People come to visit us, but we rarely visit them. In Lijiang's low tourism season, I think whether I can ask for a leave, around ten days or two weeks, so that I can be a tourist elsewhere. Nevertheless, I need a companion. Right? Here in Lijiang, a married woman cannot travel alone. If you do, others will gossip about you. This is awful. No one accompanies me to visit other Chinese cities. My husband dares not to travel, and my daughters say to me that they will once they grow up. So I wait.

A'Jie Chen's remark shows the operation of patriarchal power upon local women in Lijiang. They cannot travel, not because they do not have money, but because of a lack of companion to safeguard their moral reputation. A'Jie Chen sees many female tourists traveling alone in Lijiang, and this kind of spatial autonomy inspires her to visit other places. Nevertheless, the threat of gossip suffocates her desire for traveling to see a different world. To her, staying at home is her ultimate destiny, and the hometown has spatially fixed her into a permanent condition of geographical monogamy. By embracing the identity of hometown babies, A'Jie Chen and many other local married women bear the hidden pain of being geographically fixed in a place they are quite familiar with, but deprived of spatial mobility.

Although it is hard to resist an overshadowing patriarchal society, A'Jie Chen hopes that her daughters, the next generation of Naxi women, do not replicate her pain of being unable to travel. After graduating from vocational schools in 2019, A'Jie Chen's two daughters wanted to go to Fujian, a wealthy province on China's east coast, and learn how to professionally make tea. Their father did not allow them to go, but rather asked them to find jobs in Lijiang new city and then get married soon. In many societies, the mobility of women, whether married or not, seems to threaten a settled patriarchal order (Massey 1994; McDowell 1999). This threat was possibly noticed by A'Jie Chen's husband. Nevertheless, A'Jie Chen strongly supported their plan to go out to the coast for new skills.

> I interact with many tourists from coastal cities. They are so knowledgeable. Lijiang locals are quite conservative. That's why my husband does not want the girls to travel to and stay in coastal cities. Indeed, once my daughters get married, they have no opportunity to see the external world. It is sad that we locals rarely travel to other places as tourists or migrate. Now my daughters are still young and I want them to go out and open their horizon. They can see the difference between Lijiang and other places and learn new things. I believe that these experiences are beneficial to them.

This quote exemplifies A'Jie Chen's efforts to enrich her daughters' life experience and help them to avoid the tragedy of immobility, which happened to her. While it is hard for her daughters to evade a rigidly patriarchal society, she understands from interactions with tourists that the outside world beyond Lijiang is exciting and interesting. Although she does not have the chance to experience it, the younger generation can try a different time-space elsewhere and thus break the manacle of geographical monogamy. By morally and financially supporting her daughters to migrate to Fujian Province for career development, A'Jie Chen hopes that her daughters can choose whether to become hometown babies or stay elsewhere.

Gender relations take a particular form around A'Jie in my study. Local social norms generate a set of patriarchal relations based on the fixity of women in the domestic sphere. The grip of patriarchal control in Lijiang cannot be ignored. Local women are taught to stay at home by unwritten gender norms in rural Lijiang and the collective experiences of patriarchal power in their daily life. While local males have more autonomy to travel and migrate beyond Lijiang, local females are taught not to stay overnight in hotels, not to get married late, and not to travel alone. A differentiated geography of uneven mobilities is articulated in terms of gender to transform local women into hometown babies forever. Young (2005) summarizes that homemaking creates an antagonistic relation between man and woman, as the former's subjectivity through building and dwelling relies on the latter's objectification and dereliction, arguing that "he has a home at the expense of her homelessness, as she serves as the ground on which he builds." This sort of contrast can be applied to Lijiang. Unsurprisingly, these A'Jie demonstrate their compliance with gender norms, a compliance embodied in multiple ways in their daily life. They are bound to home and reinforce the image of hometown babies, although they may or may not dream of the opportunity to experience an alternative time-space beyond Lijiang. As a result, their self in Lijiang is embedded in localized social norms and is emotionally tied to their obligations as defined by these norms. Tourism development can lessen masculine supremacy as many local females find jobs in the tourism industry and open their horizon in the process of interacting with others. Nevertheless, they still live under the shadow of patriarchal power and bear the burden of homemaking for their family.

Having said this, I do not want to victimize these A'Jie by representing them as passive objects of patriarchal oppression in Lijiang. This representation leaves little conceptual room for uncovering the subtle ways through which these A'Jie negotiate a sense of home. For instance: A'Jie Chen encourages and supports her daughters to travel to Fujian to learn how to refine tea leaves; A'Jie Helei works hard with her husband to earn money in order to purchase a property in the new city and send their children to better schools; A'Jie Wangli maintains a good balance between serving tourists through car sharing and looking after her children. At the margins of power, these A'Jie do not complain about social injustice or patriarchal oppression, but rather actively participate in the tourism industry to negotiate asymmetrical power relations in Lijiang's local society. By casting their activities as active participation instead of sheer resistance, I attempt to articulate that these women, like other respondents I study, are part of the whole structural transformation in China. They cannot be defined as what Brown (1996, p. 732) calls "heroic soldiers in the antihegemonic struggles," but rather live with the hegemonic forces of modernization and harness their own aesthetic and monetary resources to make home for themselves and their families.

In Celebration of Hometown Babies

China's rural foundation has been eroded in the process of industrialization and urbanization since 1978. The sixth general census in 2010 showed that China's urbanization rate was 49.68%, much higher than 36.09% in 2000. By 2020, the seventh general census demonstrated that the rate jumped 63.89%, indicating that more people now live in cities and towns than in the countryside.[2] Yet despite the rapid pace of urbanization, the rural foundation is still solid in today's China, and remains vibrant in many villages. According to Fei (1992, p. 41), Chinese rural society is based on kinship and familiarity:

> People in rural China know no other life than that dictated by their own parochialism. It is a society where people live from birth to death in the same place, and where people think that this is the normal way of life. Because everyone in a village lives like that, distinctive patterns of human relationships form. Every child grows up in everyone else's eyes, and in the child's eyes everyone and everything seem ordinary and habitual. This is a society without strangers, a society based totally on the familiar.

In a spatially defined setting, this sort of familiarity results in a society of acquaintances and kinships in which individuals obtain a real sense of satisfaction and stability. This sense helps us to understand the celebration of hometown babies by Lijiang locals.

Hometown babies are rooted in Lijiang, particularly those who live close to the old town. They are deeply integrated into the tourism industry by running businesses or providing service to tourists. This integration enables them to experience a new homely lifestyle, characterized by leaving the land, but staying in situ (litu bulixiang, 离土不离乡), which means that they need not rely on agricultural production for a living, but earn money from tourism-related services. Temporally, they follow the cyclic pattern of days and seasons and enjoy a leisurely lifestyle. Their daily rhythm still corresponds to the one in traditional rural society in China: "wake with the sun and rest with the dusk" (Fei 1992, p. 57). Spatially, they revolve around their daily mobility within Lijiang valley, either for tourism-related service or social gathering. The place of their birth is the place where they grow up, get married, raise children, retire, and eventually, pass away. They may temporarily study and work elsewhere, mostly in Kunming, but choose to return to Lijiang once they feel alienated or realize they cannot firmly settle down.

In his twenties, A'Ge Yang is a driver and lives in a village close to the airport. His main job is to collaborate with guesthouses to provide chartered service between the old town and the airport. Yang does not expect to stay in China's big cities, but rather prefers to live in Lijiang and rely on its booming tourism industry for a decent living. If he is called to serve customers, then he does the job; if there is no call, he hangs out with his friends to drink and eat. To him, there is no pressure living in Lijiang, and the priority is to enjoy life. Asked whether he wants to do big business in the tourism industry, Yang admits that he has no interest, simply because the pressure and risk are too high. The preference of hometown babies does not necessarily mean that Lijiang locals are simply lazy or uncompetitive. Some choose to stay in Lijiang because of family commitments. For instance, delivery A'Ge Song, who comes from a rural village, notes his choice of staying in Lijiang: "My whole family lives in Lijiang, and the elderly stay here. It is quite unrealistic for me to work elsewhere. Easy job? There is no such thing as easiness. Express delivery is a good job, since I am satisfied with my earning. The harder you work, the more you earn."

While these above respondents never try to live elsewhere, A'Ge Gao had a different experience. He went to a vocational school in Kunming to learn how to drive forklifts and bulldozers. After graduation, he worked in different factories and mining fields. His earning was high, but he always felt lonely. Asked why he chose to return to Lijiang, he explains:

> I feel very happy to work in Lijiang. After I bring you guys to your destination, I can go home. Closeness to my own home makes me feel secure. All my friends and relatives live around. When I was in Kunming, I could not find anyone to chat with. After I went to my bedroom, it was so quiet that I felt lonely. If I had trouble, no one came to help out. Lijiang is different. If I need help, my whole clan comes and supports me.

In Kunming, A'Ge Gao was surrounded by strangers, while in Lijiang he retrieves a strong sense of familiarity so that helping each other is a social norm. Reciprocal help is not based on written contracts, but on the trust and dependability of people in a familiar society (Fei 1992). Although A'Ge Gao does not frequently seek help from his relatives and friends, he ensures that he can get it as long as he asks. His grueling experience of working in Kunming pushed him to return to Lijiang and enjoy the security and familiarity of the home village. This enjoyment affects A'Ge Gao's extended family. Except for one working in Kunming, other members, including his uncles and aunts, and younger relatives of his generation, live in the valley. To these hometown babies, staying at home not only maintains their cultural roots, but also expresses a sort of subtle distance from modernity.

Hometown babies represent a typical way of peasantry life. Different from cosmopolitans and nomads, peasants prefer to live in settled communities with security as a matter of course (Fei 1992). More often than not, this reluctance to live elsewhere is elevated into ethnic difference between Lijiang locals and Han migrants. A guesthouse owner (Siyuan, male, from Beijing) comments on the formation of "hometown baby":

> Yunnan is a unique place replete with resources. The natural environment is wonderful, and there is no need for locals to worry about food, although most of them are not rich. Automatically, they seek an easy life and feel content with the current condition. Without a strong desire for upward mobility, why don't they leave Yunnan for a challenging life? Yunnan locals reiterated that we had many business ideas. They don't want to compete with us for profit, but rather enjoy their own lifestyle. In such a social and natural environment, they feel comfortable and relaxed, and have no desire to leave the hometown.

While tourists and migrants may interpret hometown babies as conservative and lazy, I emphasize that a stay at home represents "a response to desire for fixity and for security of identity in the middle of all the movement and change" (Massey 1994, p. 151). Massey (1994) argues that individuals engage in an endless search for the real meanings of places, simply because they desperately need a bit of peace and quietness, embodied in a strong sense of place, to mitigate high mobilities and flux. This sense forges "a kind of refuge from the hubbub" (Massey 1994, p. 151). As Fei (1992) emphasizes, in traditional Chinese society, the more linkages and ties one maintains, the more intensively one is embedded in an existing social order and shows commitment to the status quo. Lijiang is no exception. As hometown babies, Lijiang locals choose to embrace small and constant changes in their own society and maintain an established time-space.

Theoretically, hometown babies represent geographical monogamy. Kinship networks born out of blood and interaction represent a dense, well-formulated structure in Chinese rural society to shape how individuals build attachment to their hometown. This is particularly true in rural Lijiang. Locals live within kinship networks that discourage them from departing to other places and geographically anchor them in a setting where they can undertake a life journey from birth to death. That is why my respondents reiterate that they can seek help from relatives and friends easily if they need it, and the popularity of huacong further expands kinship networks to localized networks through which they reinforce a sense of attachment in Lijiang. Although the rapid pace of modernization, as driven by tourism development, has profoundly changed Lijiang, this sense of attachment continues to be cherished by locals, men and women, so that they subdue the self as hometown babies and follow the rites of passage.

As Massey (1994, p. 151) reminds us, it is necessary to ask whether the relative mobility and power of one group entrenches "the spatial imprisonment of other groups." The privilege of enjoying a specific time and space by one group requires the spatial and temporal imprisonment of others, reflecting the inherent inequality embedded in social and economic development. This question is particularly valuable in the case of Lijiang, but we need more evidence to analyze whether the celebration of hometown babies by Lijiang locals and the high mobility of tourists and lifestyle migrants are interrelated. One thing is certain. It seems a win–win situation in which tourists and migrants receive services so that they can enjoy a homely lifestyle with ease in the town, while Lijiang locals can make a living from the tourism industry so that they can stay at home, instead of becoming rural-to-urban migrants in China's coastal cities. However, one explicit tension exists in this situation: the influx of information and people to Lijiang by tourists and migrants versus the continuance of home-grown culture. Within the context of liquid modernity, how can Lijiang locals maintain their roots? This challenge is noted by some respondents, as shown by A'Ge Gao's worry:

> Some dramatic changes happen to our next generation. Many kids cannot speak the native language. Kids of Bai ethnicity cannot speak Bai dialectic. Nor do those of Naxi ethnicity speak Naxi. They receive way much better education than we did, but our authentic culture gradually disappears. It is inevitable. Once Lijiang becomes a well-known tourist city, many people come to visit it. The tendency of Hanization is explicit, regardless of whether there is intervention. We have no choice.

Although Lijiang locals insist on living in their hometown, they are exposed to the inflows of information, capital, and people driven by tourism development. Gao's worry reveals that this exposure has already changed local cultural roots, as shown by the shift from ethnic dialectic to standard Mandarin. The influence of China's mainstream culture upon peripheral areas in Yunnan, or the process of Hanization, may not disrupt a sense of home cherished by A'Ge Gao and other Lijiang locals, but can dilute their geographical monogamy and draw their next generation further into a broad mobile society.

Embracing the identity of hometown babies implies both pain and joy to Lijiang locals. While we cannot argue that the current penetration of monetary capital and political power into Lijiang has undermined the established identity of hometown babies and left them disoriented, their pursuit of homely life illustrates a different way of negotiating the relationship between the self and the world. Regarding the deal of modernity, Lijiang locals choose to prioritize meaning over money. The call for being at home, as exemplified by Bhabha (1994), finds resonance with Lijiang locals who choose to be hometown babies and make efforts to maintain a settled lifestyle in a relentless world. To them, closeness to nature, and to their friends and family, is more meaningful than the race for money. While those locals close to the old town can embrace a sense of home relatively easily, those who come from distant villages are implicated in the longing for settlement between Lijiang new city and their original villages. I turn to this longing next.

Stay and Departure: The Longing for Settlement

Those who come from distant villages (two or three hours by bus, or even farther) must leave the soil as well as the home (litu you lijia, 离土又离家). Although they still stay local, their life is akin to those of rural-to-urban migrants in China's east coast. They choose to leave for Lijiang's urban area for resettlement, but still maintain social and economic connections with the ones they leave behind. What happens next? Some might spend all their savings purchasing an apartment and settle down in Lijiang's new city. This transition from rural to urban is still ongoing, but the differences related to class, education, and taste create spaces of exclusion where these new urban residents continue to feel unhomely. Some others dream of an ultimate return once they are too old to compete for jobs in the tourism market and thus choose self-retirement. Their longing for settlement somewhere implicates them in an unhomely condition in one of China's peripheral regions undergoing dramatic transformation.

A'Jie Li's life journey in the past fifteen years, as noted earlier, has swung between the old town and her marital village. In her thirties, A'Jie Li is a guesthouse cleaner of Bai ethnicity. She was born in a rural village in Lijiang and came to the old town to work in a restaurant for six years after graduating from middle school. She had a very good relationship with the boss and felt very happy. At the age of twenty, she quit the job to get married. After her first baby was born and reached the age of one, she came to the old town again and continued to work for the restaurant until it was closed. Then she returned to her marital village and had her second child in 2016. In 2017, she left the village for the old town and worked as an A'Jie in a guesthouse. In 2019, the guesthouse closed, and A'Jie Li was recruited to work in another guesthouse. Several months later, she had to quit the job and stay at the village to take care of her two children. Her husband runs a small construction team in Lijiang's new city so that A'Jie Li must stay in her marital village. As she notes, "both parents cannot be away from our children. Otherwise they won't study hard. At least one adult must be at home and manage things such as childcare, raising pigs and goats, and looking after the elderly." The COVID-19 pandemic hit Lijiang's economy very hard, and her husband cannot earn adequate money. Realizing the need for more money for their children, A'Jie Li left the village once the pandemic waned in April 2020 and continued to work as a cleaner in the previous guesthouse.

A'Jie Li visited the village two or three times per month and two days per visit, simply because she deeply misses her two children. She explains: "I am used to the separation now. At the very beginning, I missed my children very much and even cried after I left them for the old town. What a sad experience! But I have no choice. When they grow up, their education costs will be quite high. We need to earn money to support them." Regarding the future of her life, A'Jie Li dreams to buy an apartment in Lijiang's new city so that she can relocate her children from the rural town for better education. This investment means a huge burden and challenge to A'Jie Li and her husband. If her son comes to Lijiang's new city for middle school in 2022, A'Jie Li will quit her job to look after him and ensure his academic excellence. She also plans to bring her daughter to the new city to attend a primary school. Then the whole family will reunite and start a new life.

Compared with rural life in her marital village, A'Jie Li is quite satisfied with the urban life in Lijiang and reiterates that it is impossible for her family to settle down in the village anymore. Asked why, she explains:

> Women of Bai ethnicity are very hardworking. In the past, we did all the domestic work and men went out for money. My in-laws help me to look after my children. There is no hope to rely on farming to make a good living. My family's land size is very limited. I feel very happy to work in the old town, although the salary is not high. It can easily run out if we go shopping for clothes.

As the village has no hope for livelihood improvement, urban Lijiang becomes the only place with the possibility of a better life in the future. Having said this, A'Jie Li's longing for settlement in urban Lijiang faces enormous challenges. One key impediment is the skyrocketing cost of housing. Due to rapid tourism development, Lijiang's real estate market has undergone dramatic expansion. The average housing price in Lijiang's new city in July 2021 was CNY 8,400 (~US$ 1,239) per square meter.[3] If A'Jie Li wants to buy a 100 m^2 apartment, the total price is about CNY 840,000 (~US$ 123,900). In 2020, the average disposable income per capita in urban Lijiang was CNY 42,910 (~US$ 6,329) (Bureau of Statistics in Lijiang Old Town District 2021). This gap between income and housing price is clearly felt by A'Jie Li and her family, as she reiterates that it is extremely difficult to purchase a property in Lijiang new city.

A'Jie Li's life trajectory is quite typical among others I interviewed in Lijiang. Their constant journey between marital villages and urban Lijiang places them in a real swing between here and there, between the family home that they leave behind, but cannot abandon, and the dream home that they cannot afford to secure in urban Lijiang. In another case, A'Jie Helei and her family moved to urban Lijiang and work for the tourism industry. What reminds her of the previous home is the farmland allocated to her and her husband, but the farmland, together with her old house, is unattended and thus almost deserted. They occasionally return to the village, only for important family affairs. In urban Lijiang, they no longer engage in farming and tobacco plantation. Rather, all of her family members' lives revolve around the tourism industry, which they earn money from tourists directly or indirectly. In her old house the whole family lived together under the same roof; now family members are scattered around and need to find time to gather together. Even her husband often stays overnight in other places as required by long-haul tourist journeys.

A'Jie Helei's new home in urban Lijiang is loosely bound not through daily life or an apartment, but by emotional ties among family members. Therefore, A'Jie Helei has undertaken her own great transformation: from agricultural production to tourism service, from rural village to urban area, and from a close-knit family to a loose-knit one. She admits that she cannot return to her previous lifestyle. Asked why, she notes that

> I can probably return to the village. After spending five years in the city, I cannot bear the weather in the village. It is too hot. In the old town, I work inside bedrooms. In the village, I had to work under the sun. Yes, I have to say that I am not used to the rural life. We had a lot of land, but left it deserted.

As her whole family can now unite together, A'Jie Helei feels a sense of home in Lijiang's new city. Although she owns no property nor has extensive social networks in the new city, she still longs for settlement and becomes an urban resident. The old home that she and her husband created in her martial village became a place to which she could not return, but the new home exits only for the reunion of her core family.

While both A'Jie Li and A'Jie Helei specify the impossibility to returning to their marital villages, other locals still cherish the hope of a return to their rural villages in the future. For instance, delivery A'Ge Heli has a clear plan for him and his wife. He affirms that they need to save money to furnish their rural house and purchase new furniture in Yuhu village. Possibly after five years, they will leave the urban area and return to their familiar rural lifestyle in Yuhu. As Heli notes, "Several years from now I will be older. If I cannot ride a tricycle to cart luggage for tourists, I will go home (huijia, 回家)." When asked what to do in Yuhu, Heli considers running a family restaurant to cater to the city's residents for huacong or to tourists who want to try local food. A'Ge Heli rarely regards himself as a typical rural-to-urban migrant, but rather a local looking for business opportunity around his hometown. When talking about home, A'Ge Heli and other service workers have a clear geographical reference point – their hometown or village – to which they plan to return one day. In other words, they pursue a synthesis of local and translocal elements to make their home. Many of them feel at home in urban Lijiang because they think they are locals and live an ordinary life, but have a plan to return to their hometown as long as the moment is coming. Different from round-the-world travelers who develop a cosmopolitan claim to the world as a home by encountering difference, but always just passing through (Germann Molz 2008), service workers in Lijiang make efforts to make their home in the new city, but feel unsure about when and where to settle down permanently.

The temporal and spatial reorientation towards the home(town) they left behind has been captured by Schlecker (2005) in his analysis of urban residents in Hanoi and their rejuvenated socioeconomic ties with the home place, that is, their ancestral villages where their parents or grandparents originally came from. As Schlecker (2005, p. 520) emphasizes, these residents regard the home village as "an alternative image of belonging and relatedness where the past offers certainty for relationships in the present." Here, hilarious gatherings in the home village revitalize temporal and spatial connections through which the past and the remote are celebrated and cherished by way of gift-giving and mutual greetings. Schlecker's writing indicates that in a rapidly changing society, the past gains more importance than ever by generating reference points of stability and certainty for people to maintain their dramatic life experiences. The value of an old hometown (laojia, 老家)is to maintain "the myth of return" (Kochan 2016, p. 25). Sustained by this myth, rural migrants in cities think that they have a Plan B if they must eventually uproot their adventure from one urban compound to another (Zhan 2022). As Kochan's (2016, p. 25) analysis of rural migrants in Beijing and Shenzhen shows, the laojia does not merely entail a socially stable home, but more importantly, supplies "a conceptual and practical safety net that enables migrants to explore other socio-spatial possibilities in the city."

Nevertheless, we must be careful about the possibility of such a return. Indeed, more and more people choose to leave rural villages for jobs in urban Lijiang. When numerous villages are hollowed out and even abandoned as a result of massive migration, these service workers will have difficulty going home, as their home in villages, the embodiment of security and familiarity, will be gone forever. A'Jie Li notes the dramatic demographic change in her marital village:

> People of my age in the marital village almost all leave it to work in urban Lijiang. They don't stay in the village, but only return for some important village affairs. Some of them purchase apartments in Lijiang's new city and settle down as urban residents. Those who still stay in the village are elderly persons. Some couples also leave their children to be cared for by the elderly in the village.

The more distant the village is from urban Lijiang, the more dramatic this demographic change. Whether they want it or not, these service workers from rural Lijiang remain sojourners in the city, or dwell with mobilities. Hence, a tale of "two Chinas," as Huang (2008) and Wong et al. (2022) find in the cases of modern cities and semi-modern and traditional villages, is replaced by multiple Chinas with endless mobile flows within and beyond the country.

Different from other service workers I interviewed, A'Jie Yue faces real challenges to engage in a settled life. In her forties, A'Jie Yue comes from Heqing, a county 50 km from Lijiang Old Town. Administratively, Heqing is not part of the prefecture of Lijiang, but is close enough to the old town that thousands of residents in Heqing have migrated since the late 1990s to urban Lijiang for business and job opportunities. In 2014, A'Jie Yue's husband died of a disease and left her a debt of CNY 70,000 (~US$ 10,320). Without a choice, she had to earn money to pay off the debt. Her sister introduced her to a guesthouse in Lijiang Old Town. Since then, she has stayed in the old town and worked as a cleaner in various guesthouses. In the high tourism season, A'Jie Yue is very busy, as she has multiple jobs. By 2018, she paid off the debt. In the past, she had to return to her marital and maiden villages in Heqing to handle family matters, including organizing her in-laws' funerals and looking after her sick parents. Thus, she had to ask for a leave, and her guesthouse manager sympathized with her challenges and approved her leave without pay. At the time of our interview in 2020, her younger daughter was in a high school in Heqing and visited A'Jie Yue every week, while her older daughter worked in a beauty salon in Guangzhou.

Although she resides in the guesthouse where she works, she still rents a bedroom in an urban village near the old town. When her younger daughter comes to visit her, they stay together in the rented bedroom as a family. To A'Jie Yue, her daughter's weekly visit makes her feel happy and consolidates her sense of home. In the neighborhood, many residents come from her village in Heqing county and work in Lijiang's tourism industry. They reestablish social ties so that a kin-based village culture is created to replicate the one they left behind, but remain detached from the land and rural lifestyle. While her fellow villagers, particularly the females, love to dance together for socialization and exercise, A'Jie Yue prefers to watch them or rest in her own room. Once they are integrated into Lijiang's tourism industry to conduct service-oriented jobs, their established ties with the land disappear and new social structures are formed, a process that makes these villagers feel uncertain and even lost. The short distance between Lijiang Old Town and their "home" village, 50–60 km, generates a hope that one day they can return if they are tired of urban life or cannot make a living in urban Lijiang.

To A'Jie Yue, however, there is no such thing as a return. As she explains, "Although I was born in Heqing, I think now I become a half Lijiang person. I cannot return to my marital village [in Heqing]. After spending years in Lijiang, there is no way for me to return. I have no land in the village. And my in-laws have passed away. It is meaningless if I return." Ever since her husband passed away and her two daughters left the village to follow their own pursuits, A'Jie Yue's spatial and social connections with the place that she once called home evaporated. As a married woman, there is no reason for her to be reintegrated into her maiden family. In China's rural society, a woman who has been married to her husband is like water that has been splashed. Unsurprisingly, A'Jie Yue calls Lijiang a home, although this home is temporary.

Regarding the future, A'Jie Yue has no plan. Once her younger daughter is admitted to a university, she hopes to migrate to Guangzhou where her older daughter works and to find a job with higher pay. Nevertheless, she is much worried about the high cost of living in Guangzhou. If she continues to stay in Lijiang, she cannot help but miss her daughters. To A'Jie Yue, her home is not anchored in the abandoned house in her marital village. Nor does she regard the guesthouse where she works as a home. There is no place that A'Jie Yue can call home; only her two daughters generate a sense of home to her. Different from other respondents I encountered in Lijiang, A'Jie Yue has not actually engaged in the deal of modernity. But her unhomely life comes from the breakdown of her family and her life experience of temporary home in urban Lijiang.

A'Jie Yue's temporary home is different from houselessness. She does not live on the street, nor is she jobless. Rather, she maintains decent places to dwell in the old town as well as the new city, and works in two guesthouses. She temporarily works and resides in the old town, but her future depends on her two daughters. Being temporary entails two interrelated components. First, it indicates a temporary stay in one place and a motion of traveling to other places, and second, it involves a shelter to make dwelling possible. In this dual meaning as both traveling and dwelling, a temporary home should be understood as an in-between placement in constant flux. A temporary home is fluid here and now, but embodies a sense of belonging and attachment, or what Ahmed (2000, p. 88) calls "complex and contingent spaces of inhabitance." This complex and contingent nature calls for more attention to the blurred boundary between mobility and mooring. A'Jie Yue's sense of temporary home in Lijiang reveals dwelling and traveling as two constituent features of liquid modernity (Bauman 2000).

Writing about rural-to-urban migrants in Shenzhen and Beijing, Kochan (2016) explores the evolution of their conceptualization of home through three lenses – the hometown (laojia) they left behind, the city as a home, and the material space of dwelling. Similarly, Tang et al. (2022) found that rural migrant women in China negotiate the patriarchal norms embedded in Confucian ideology to build their lived and idealized home at multiple localities – laojia and big cities. In cities, these women can escape from persistent gender discrimination expressed in rural areas and develop a sense of home based not on property ownership, but on independence, respect, and engagement in urban life (see also Pun 2005). Here, local service workers from distant villages struggle between here and there, and this struggle is differentiated by age. Those in their fifties or older visit original villages to participate in significant events such as weddings and funerals, or to handle clan affairs. Those in their twenties and thirties pay a return visit to show care for their elderly parents or reunite with left-behind children. While the former develop a hope of return once they are tired of hard work, the latter long for resettlement in urban Lijiang for better infrastructure and education. In his study of mine workers migrating from the rural countryside in Zambia, Ferguson (1992, p. 81) notices the existence of unreturnable home for these workers, arguing that they regard "the 'home' village as a faraway place, rarely if ever actually seen or visited, to which one is connected more by nostalgia and sentimental attachment than by social and economic ties or life trajectory." This unreturnable home happens to those young service workers in urban Lijiang, but the skyrocketing price of housing in the new city poses enormous challenges to them for permanent settlement, and thus jeopardizes their ability to achieve a settled life. Between an unreturnable home in rural villages and an unsettled home in urban Lijiang, these young service workers must endure unhomely life for a while. To those elderly, however, going home to their original village represents a permanent longing for settlement, or for a return to the soil where they can eventually rest their soul.

Conclusion

Hometown babies represent a typical way of peasantry life in China. In 1946, Fei (1946) found that different from cosmopolitans and nomads, peasants lived in settled communities with stability. Kinship networks born out of blood and interaction constitute a dense, well-formulated structure in Chinese rural society to shape how individuals build attachment to their hometown. This sense of settlement is particularly strong in the Lijiang valley, a sense that discourages them from departing to other places and geographically anchors them in a setting where they can launch a life journey from birth to death. This strong home culture in Yunnan in general and Lijiang in particular makes people conform to established social norms so that they can maintain the stability and continuity of the society. Bereft of risk taking and high mobility, the society itself remains coherent and structured. The home culture limits Lijiang locals' ability to adapt to change in periods of social transition brought by tourism development. Ironically, this culture generates a fertile soil to foster Lijiang's selling point – loafing and relaxation – in the tourism industry. Lifestyle migrants and retreat tourists come to visit Lijiang to remedy a loss of home in their hectic life, and work together to bolster a sense of familiarity and solidarity, the kernel of Chinese rural society, through leisurely activities such as drinking tea and having dinner in guesthouses. Today, local people in Lijiang maintain their own lifestyle to withstand "the unreality of the modern world" and become hometown babies forever. This lifestyle is lost in China's big cities and thus attracts middle-class people from these cities to travel or migrate to Lijiang to experience a sense of home characterized by closeness to nature and a slow pace.

The lingering sense of hometown babies represents a continuity of place attachment in the swirl of China's great transformation. After spending eight years in the Lijiang valley in the 1940s, the Russian-born traveler Peter Goullart (1955) noticed that the concept of time in Lijiang was totally different from that in the West, arguing that the greater part of time in Europe and especially in the United States was devoted to making money to such an extent that individuals were preoccupied with work and the ritual killing of time. Goullart (1955, p. 183) attributed this obsession with work and busyness to "the unreality of the modern world which man has created for himself." In Lijiang, however, Goullart (1955, p. 184) observed that "[t]he people in the street interrupted their bargaining to admire a clump of roses or peer for a minute into the clear depths of a stream. Farmers paused in their fields to gaze at the ever-changing face of the [Yulong] Snow Mountain." Without being touched by the complexities and hurry of modern life, time in Lijiang was not for money making, but for meaning making. As a result, Naxi people rarely wanted to leave the valley. Even for those who had seen the materialistic glories of Shanghai, Hong Kong, and Calcutta, they still preferred to return to Lijiang to live. This preference is captured by Fei (1946) who emphasizes that people in *xiangtu* society often intend to stay in one place forever and become geographically monogamous (Heller 1955). Today, the tempo of life highlighted by Goullart continues to exist in the Lijiang valley, and the reluctance to leave Lijiang still characterizes local people.

Nevertheless, this cultural continuity unfolds differently among different groups of people. Service workers from Lijiang's rural vicinity leave their original villages for the old town and work hard to make the town a comfortable and convenient home for migrants and tourists. While migrants and tourists have been enjoying the town as an "ideal" home for loafing and simplicity, local service workers hurriedly fulfill their cleaning or delivery duties. Those service workers from distant villages in Lijiang not only lose their intimate connections with their original community, but also become the object of exploitation and discrimination to such an extent that the promise of modernity they had been offered in the course of economic reform turns out to be a real challenge. They are living in between the city and the village, and this in-between position challenges their root and attachment. They are thrown into the orbit of marketization and urbanization, actively or passively, and get immersed in a space produced by state power and market forces, in which their autonomy is stripped as long as they strive for a living. To those who come from distant villages, the old town is merely a temporary stop in their life journey because their symbolic home is located in the villages they left behind. The longer they stay in Lijiang Old Town, the more they get used to an urban society imbued with people, information, and for sure, capital, and the harder it is to depart for their original home in the countryside. Nevertheless, high housing prices in the new city suppress their longing for settlement and push them elsewhere. Struggling between stay and departure, they live an unhomely life.

Those service workers from the old town's vicinity have more ability to enjoy a settled life. By providing services to tourists, they can earn money without leaving their home villages. Although they are part of the tourism industry, their time and space is outside the spatial and temporal order of the town. They acknowledge that this order implicates them in the process of tourism production and consumption and has impact on their life, an order that they even find contrasting to their regular routines, but they keep a distance from the rat race for wealth so that they can have more time to stay at home and enjoy a homely life. They realize that they cannot compete with migrants for more profit in the tourism industry because of a lack of monetary capital, management skills, and entrepreneurial ideas. Neither do they criticize the system of tourism development, market competition, and the fast pace of life elsewhere. Nevertheless, they feel quite satisfied with their well-established patterns of life – slow and relaxed – supported by marginal benefits from tourism development such as rent, small business, and service. While speed is an inevitable order, totalizing and irreversible, to implicate Chinese cities in an endless race for wealth and fame, Lijiang locals maintain a distance from speedy modernization and refuse to make substantial changes. A rat race for wealth and fame hits an invisible wall in Lijiang, and this wall becomes lucrative to many, thanks to the lingering desire for a sense of home. Nevertheless, Lijiang locals are worried about the penetration of modern lifestyles into their neighborhood, as increasing numbers of young Naxi locals have been deeply integrated into a consumerist society at large. As they do not rely on agricultural plantation for a living, their established attachment to the soil, which is regarded by Fei (1946) as one of the characteristics of rural society, is in peril.

Although the rapid pace of modernization driven by tourism development has profoundly changed rural Lijiang, these social norms cast a shadow upon locals, both men and women, so that they subdue the self as hometown babies and follow the rites of passage. These norms are mostly verbal, and even outside the legal framework, but Lijiang locals know their power and understand that their morality depends on compliance with these norms, rather than on a rule of law pertaining to the state. Tourists travel to Lijiang Old Town because they want to; service workers travel to the old town because they have to. Relaxation, enjoyment, and loafing constitute the essence of tourism consumption in Lijiang, and the experience industry is geared to tourists' desire for spatial and temporal autonomy. The fulfillment of this autonomy requires, however, the arduous contributions made by service workers. The side effect of loafing and relaxation in Lijiang is the transformation of guesthouse cleaners and delivery workers into disciplined subjects who have no time to relax and slow down. These workers are, one might say, essential to making Lijiang Old Town hospitable and homely, but they remain unable to claim the town as their home.

Notes

1 In this chapter, the majority of locals that I study also possess Lijiang household registration status. A fraction of them come from Heqing, a county 50 km from Lijiang Old Town. Albeit an administrative part of Dali Bai Autonomous Prefecture, Heqing has established a close connection with Lijiang simply because Heqing people have run businesses in Lijiang Old Town since the Tea Horse Road period. The Heqing locals I study in this chapter do not hold Lijiang household registration status, but they have developed a cultural affinity with Lijiang.

2 https://china.huanqiu.com/article/44uUfgF1N1u

3 https://lj.fang.anjuke.com/fangjia/guchengqu/?from=sub

Chapter Seven
Homemaking in a Relentless World

I began *Unhomely Life* with a story about a domestic lifestyle migrant, Mahua, who felt homeless in Beijing and was reluctant to return to her hometown in Jiangsu Province, but chose to temporarily stay in Lijiang Old Town. Immersed in a life journey from one home to another, I set the stage for my argument that home remains a focal point for the critical analysis of continuity and disruption in China's great transformation from a Maoist planning economy to a rising global power in four decades. Within the context of speedy modernization, Mahua's sense of homelessness is not exceptional once capitalist modernity brings about a disorientation of the qualitative meanings found in time and space towards purely quantitative rationality, as embodied in the abstraction of commodity-based exchange (Gardiner 1995). Time and space should be used efficiently for mass production and profit making, and any alternative forms of use, including loafing and laziness, should be abandoned and even disciplined. Once time and space are subjected to a purely bureaucratic structure of rationality to facilitate the logic of capital accumulation, modernity results in a loss of meaning and depth. To be modern is to be part of a universe in which, as Marx said, "all that is solid melts into air."

The history of capitalism is a history of home unmaking. Efforts have been made by political and economic forces to coerce individuals away from the elements of nature, first, the soil-based natural environment of countryside and then, the natural cycle of time. When modernization threatens to afflict individuals with a homeless mind for the sake of capital accumulation, the home could be a solid resort to counter this inevitable melting process. The desire to embrace the

Unhomely Life: Modernity, Mobilities, and the Making of Home in China, First Edition. Xiaobo Su.

natural environment and cyclic time, which echoes what ancient Chinese scholars define home – a destination of return and a balanced way of living, is one among many activities for individuals to imagine and experience home, either temporarily or perennially, to retrieve the meanings of space and time, and thus rest their soul for inner calm. As I have emphasized throughout this book, homemaking is an important element in orchestrating the time-space of daily life, which must be reflected in a configuration of critical geographies of modernity. Apparently, the relationship between home and homelessness in the history of modernity cannot be understood as antagonistic or linear. Rather, in a peculiar way, these two conditions follow more closely the dialectical evolvement, in a spiral dynamic, to advance the life journey from there to here, from past to present. The longing for home generates the imaginations and experiences of sensed home in accordance with respective cultural values in order to counter the ubiquitous sense of homelessness deduced by speedy modernization. We need to direct our attention to the interrelationship of home and homelessness in a relentless world.

The deal of modernity between money and meaning, as discussed in Chapter 1, seems to oversimplify human choice in relation to the representations and experiences of home. Having said this, this deal constitutes an important angle from which we can analyze how people are able to demand improvements in the context of social acceleration. The simplification of such a deal should not, however, mean a simple analysis of the complex relationships involved in the process of homemaking. Indeed, concepts such as home, homelessness, unhomely life, and even modernity are not neutral categories of empirical analysis and theoretical thinking, nor do they simply reflect European experiences following the Industrial Revolution. These are key modalities to reflect human nature in the current stage of social and economic development in relation to capitalist modernization. When tackling such concepts, our job is thus not merely to refer to Europe, but to investigate how these concepts are embedded in different geographical contexts to exemplify the consequences of homemaking and unmaking.

All these debates – at the most epistemological level – deal with how individuals mediate with and adjust to speedy modernization in China and elsewhere. The ability to feel at home, which represents the agency to counterbalance the spatial and temporal changes brought by modernization, falls differentially in these subjects' everyday life. Furthermore, unhomely life unfolds between dwelling and traveling, here and there, past and present, to encompass and embody the multiple and even ambivalent experiences through which people engage in an endless journey of homemaking. It is the in-between condition – home and homelessness – that excites me to explore the variegated forms of negotiations for sensed home. The hierarchies of power in terms of class, gender, ethnicity, and age coalesce to shape how people make and unmake homes for the purpose of money or meaning. Continuity and disruption permeate China's structural transformation from a *xiangtu* society to an unhomely society in the post-Mao era, and individuals use their variegated abilities to adjust to this transformation. By treating Lijiang Old Town as a contact zone where mobile groups of people converge, we find a nuanced picture of how these individuals struggle against homelessness and endeavor to make home. Regardless of their socioeconomic backgrounds, my respondents' interactions with and in Lijiang demonstrate the central importance of home as a source of comfort and belonging by which they can live with speedy modernization in China's first-tier cities such as Beijing and Shanghai, as well as small cities such as Lijiang on the periphery. By swinging between home and homelessness, they hope to pursue an acceptable and balanced way of life in a dramatically changing society.

The Politics of Homemaking in Lijiang

The imagination and experience of home in Lijiang is part of a larger pattern of structural transformation prominent in China. As Chinese President Xi Jinping emphasizes, China has entered a new era of socialism, a time when China's principal contradiction has evolved into one "between unbalanced and inadequate development and the people's ever-growing needs for a better life."[1] There is no doubt that many people in China are still struggling for food, but many more can transcend pure materialistic gain to pursue a better life. Like people who seek therapeutic consultation to ease anxiety in China's big cities (Zhang 2020), the individuals I study converge in Lijiang to fashion a milieu different from hectic cities so that they can negotiate the deal of modernity in the process of making and unmaking home. To them, no home is permanent; all homes in Lijiang are made for placement and inhabitation, but are in constant flux. In such a tourist destination where people come and go, their homes become mobile and even temporary, endowed with "complex networks by which 'hosts, guests, buildings, objects and machines' are contingently brought together to produce certain performances in certain places at certain times" (Hannam, Sheller, and Urry 2006, p. 13).

As modernization and urbanization continue apace, middle classes in China develop a market taste for loafing and simplicity. Lijiang Old Town is particularly well-known as it generates a slow lifestyle for tourists to get away from busyness and acceleration. With the mix of transitional Naxi ethnic culture and a comfortable natural environment, in contrast to a specifically built destination with various attractions, Lijiang can be better understood as offering tourists and migrants an entry into a variety of opportunities to conduct self-reflection and enjoy loafing, far removed from the hustle and bustle of urban life in China's big cities. In a period of social acceleration, such a destination thus comes to be understood as standing not just apart from modernity, but in some sense beyond modernity as the spirit of time-discipline sweeps across Chinese metropolises such as Beijing, Shanghai, and Guangzhou. The pursuit of an alternative lifestyle by retreat tourists and lifestyle migrants neither requires locals – native residents or service workers – to be insulated from modernity, nor calls for an authentic Naxi town by preventing the town's commodification. Rather, the development of a home-oriented experience economy draws on a mixture of Naxi established style of loafing, natural environment, and invented stories.

To lifestyle migrants, a long stay in Lijiang is imbued with patterns of feelings, including hope for healing, desire for relaxation, and an idea of escape. They do not characterize their preference for loafing in Lijiang as a new form of hedonism that characterizes contemporary consumerism. Rather, they migrate for a different aura, akin to China's *xiangtu* society, to retrieve the meanings of time and space. Perhaps loafing in Lijiang represents a return to a traditional lifestyle lost in the process of modernization in the past four decades, but it happens only in a town where modern infrastructure is available, daily chores are done by local service workers, and risks and troubles minimized by the Lijiang Heritage Authority. The idealization of Lijiang Old Town does not mean to present an illusion of a last horizon unspoiled by the ills of modernity. Rather, these migrants attempt to strike a balance between money and meaning, that is, enjoying their favored lifestyle, which is different from social acceleration, and selling this lifestyle to tourists to enable them to maintain this more relaxed lifestyle. In a small town suffused with economic exploitation, social stratification, and business stress, these migrants make efforts to maintain their spatial and temporal autonomy and enjoy a sense of home, no matter how long they can do so.

For retreat tourists, a temporary stay in Lijiang means self-rejuvenation and intentional adjustment so that they can better handle hectic urban life once they return to the big cities. Traveling to Lijiang or other similar destinations appears to be a comforting and acceptable solution to the loss of home in Beijing, Shanghai, Shenzhen, and other highly modernized Chinese cities. Their temporary stay entails a "time out" from their usual lifestyle imbued with stress and speed, as well as a "place out" through a spatial retreat into Lijiang for different surroundings, including people, natural environment, and time. Or, if Lijiang represents a utopia, this utopia is reachable and consumable by their purchasing power. A retreat into Lijiang spells out an acceptable escapism, as my respondents cannot bear the real loss if they were to cut off all economic and social ties with their established wealth and fame. They must return to their abode in big cities and resume their overloaded life so that they can earn adequate money and time off for their next trip to Lijiang or elsewhere.

If the overriding force in an era of capitalist modernity is the annihilation of space by time (Harvey 1982), the longing for home is the refreshment of time by space, that is, traveling elsewhere to experience a different tempo. Nevertheless, we must be cautious about how much the Lijiang lifestyle can offer magical solutions to the ills of modernity in China. Furthermore, the home industry in Lijiang Old Town pushes native residents into an uncanny condition of feeling strange in their own hometown. In this condition, these residents must handle in-situ displacement, which induces a mixture of feelings, such as anxiety and anger about the town's inevitable transformation into public space, and nostalgia for and enjoyment of their own ancestral houses. Many native residents rapidly moved out to the new city either to receive monetary return from house rents, or because of uninhabitable living conditions in the town, or both. Those who remain feel that their established social fabric has been dismantled and recognize that the town is not their home anymore. Once their rooted home gives way to a home industry constituted by migrants and tourists, native residents feel that they become unhomely, whether they choose to depart to the new city or stay in the old town.

The experiences of a homely lifestyle in the town by tourists and migrants are made possible by service workers. Tourism development in Lijiang brings job opportunities to local people and helps them to maintain the identity of hometown babies. For those locals who live close to the old town, their insistence on a well-established lifestyle in the valley might appear to contradict the sense of progress and advancement that pervades the trajectory of China's modernization. Yet it offers a glimpse of how the meaning of life in time and space, even under conditions of economic growth, receives priority in the communities in rural Lijiang. By maintaining and enjoying their own lifestyle, this continuity of homemaking opens up avenues of social stability and place attachment for Lijiang locals to withstand the economic and political visions of modernization. For those who come from distance villages, they must leave their home to ensure the smooth operation of the home industry in Lijiang Old Town. Close-knit community ties continue to exist in rural Lijiang and shape service workers' mobilities between the villages they leave behind and the new city where they struggle to settle down. They still recognize kinship and return to the village on ceremonial occasions. But these A'Ge and A'Jie face different challenges and opportunities of homemaking.

The comparison between tourists/migrants and natives/locals illuminates the inherent contradiction between here and there. Generally, *xiangtu* society is unattainable, but can be experienced as a tourist destination by those who stay in modern cities and pay for a nostalgic taste of time and space. This nostalgia for premodernity, as Tinsley (2020) emphasizes, remains an inextricable component of the European imperial project. Ironically, those citizens in premodern societies dream of modern facilities and technologies in industrial societies and take arduous cross-border journeys to the hearth of modernity to remake their new home. The former's nostalgia for premodernity juxtaposes with the latter's yearning for modernity, implicating both in an unhomely condition. This entanglement similarly happens in Lijiang, albeit in a less violent manner. Everyone remains mobile or dislocated; very few can develop a settled way of life in Lijiang Old Town. Thus, the unhomely constitutes a fundamental, inherent element of daily experience in the course of China's great transformation, to epitomize the uncertainties and struggles of modernity itself.

What is important to note in Lijiang is that tourism and migration are always embedded in the broad politics of uneven mobilities that revolve around questions of who can move, when, how, and for what purposes. While retreat tourists visit Lijiang to slow down, spend money, and waste time, rural villagers leave for urban Lijiang to speed up for money. While female Han tourists can ignore gossip and stay in Lijiang as long as they want, local Naxi females are spatially implicated in Lijiang to fulfill the role of homemaking. While Lijiang locals prefer to stay static and enjoy being hometown babies for a relaxed and slow lifestyle in the valley, lifestyle migrants are torn apart between a new home where they feel uprooted and an old home that they leave behind. While elderly natives struggle to imagine a sense of community and stay in their ancestral courtyard houses for inner calm, young people rush to move out to the new city for a taste of modern styles. This comparison can go on, but one thing is certain. Voluntary mobility is "a social good, a resource, not equally available to all" (Skeggs 2004, p. 50). Or as Massey (1994, p. 151) critically states, "We need to ask, in other words, whether our relative mobility and power over mobility and communication entrenches the spatial imprisonment of other groups."

The use of time adds one more dimension to uneven mobilities in Lijiang. Delivery A'Ge helps clients – business owners and tourists – in the town to free up time for leisure and enjoyment, but they simultaneously trade their time for salary, and are implicated in hectic and precarious working conditions. The contrasting temporalities between Delivery A'Ge and clients exemplify the multiplicity of time in orchestrating the practices of life – as a means of production, a way of living, and a form of value. The divergence of temporal experiences further enriches the meanings of the town to different users. To A'Ge, the town is a bustling site where they must rush from one client to another; for lifestyle migrants, the town is a space of swift transaction and slow relaxation; for leisure tourists, the town is a modern utopia to counter the acceleration of the everyday so that they can better rejuvenate their body and mind for hectic urban life in their regular abode; and to displaced native residents, the town is a symbol of the past lost in its transition from a vernacular place for daily living to a highly commercialized theme park for property rent. This comparison highlights that time is not simply the background against which human activities take place, but also that time structures daily experiences and is structured through human practices (Besedovsky et al. 2019). In one place such as Lijiang Old Town, multiple temporalities coexist to foster variegated meanings of space.

Patriarchal orders pervade daily life in China, but affect women in different ways. To Lijiang locals, gender norms are attached to homemaking so that women are discouraged from travel or migration. To female tourists traveling alone from big cities, they derive critical insights from their embodied experiences of loafing in Lijiang and draw on these experiences to ignore gossips and formulate new subjectivity related to independent selfhood. As these two groups of women converge in Lijiang, we can find a multivalent enmeshment of struggles against and compliance with established gender norms in Chinese society. Tourism development in the town pushes them to rethink other possibilities of life. Lijiang local females work hard to provide service to tourists, but engagement with tourists allows them to earn money, expand their social horizons on the world, and thus develop new possibilities of enriching themselves. Meanwhile, those female tourists from big cities visit Lijiang for a sense of home to such an extent that they can enjoy spatial and temporal autonomy, which is made possible by female service workers. Uneven mobilities reveal how class and gender coalesce in embodied forms. In the space of tourism production and consumption, gender norms constrain Lijiang local females who come to stand in for geographical mooring, but become the target of transgression by those female tourists from big cities.

Nevertheless, we must be careful about a binary thinking of these dynamic relationships among different groups of mobile people. It is not necessary to divide them into the powerful and the powerless. As I show in *Unhomely Life*, Lijiang entails multiple systems of hierarchy in terms of gender, class, age, and ethnicity. When different systems interact, individuals can be simultaneously powerful and weak. For instance, lifestyle migrants own more cultural capital and entrepreneurial skills than local property owners, but become quite weak when they have economic disputes with the latter. Both retreat tourists and lifestyle migrants, who seem to enjoy voluntary mobility, stay in Lijiang to feel placed or rooted, but they cannot escape from a lingering sense of outsideness or alienation. Neither can they stay away from the specter of homeless mind born out of rapid modernization in China's big cities. At the same time, Lijiang locals choose to embrace and even maintain their established lifestyles, which unintentionally help them to mediate the encroachment of modernization into their daily life. Being a hometown baby is not a shame. Similarly, being a cosmopolitan may indicate a privilege, but the price is a permanent search for home, which always exists elsewhere and elsewhen.

Uneven mobilities should be rethought to shed new light on the dialectical relation between privilege and deprivation. On a micro scale, these individuals are empowered and disempowered in the endless interactions with others in the production and consumption of tourism. From a macro scale, all my respondents revealed their struggle to pursue and maintain a homely life in the course of China's speedy modernization. Having said this, I intend not to make these individuals victims through representing them as the passive objects of modernization and urbanization in China. By analyzing their efforts to enjoy a homely lifestyle and feel at home, however, *Unhomely Life* unpacks the subtle and ambivalent ways in which they deploy the imaginations and experiences of home, cherished in Chinese culture, to negotiate time discipline and sociospatial alienation so that they can resist and conform to the high rationality of modernization.

Loss of home feelings and the search for home becomes central to China's great transformation in the post-Mao era, contradictory and intertwined as that is, and forges an intricate dialectic to shape individuals' life journeys. Speedy modernization in China is suffused with the loss of home feelings by uprooting individuals from their home in order to get rich first and fast. At the same time, people have more purchasing power to pursue and experience a homely way of life to mitigate the ubiquitous existence of homelessness and uprootedness. This dialectic, then, is a central theme of this book: the more modernization lifts people out of *xiangtu* society or soil-bound society, the harder they strive for a sense of home embedded in *xiangtu* society. Certainly, a loss of home feelings is not the only cultural form of China's great transformation. Mental stress, conspicuous consumption, movies, music, and other cultural forms can, I believe, be meaningfully investigated for this crucial relation between modernization and everyday life. The making and unmaking of home, however, touch upon the foundation of Chinese society and give human agency particularly far-reaching clout in a rapidly changing context. By unpacking the struggles and enjoyments in everyday manifestation of making and unmaking home, I want to emphasize that these reactions are not unique to China, since what happened in China has already taken place in many other geographical contexts since the Industrial Revolution. The specter of homelessness continues to haunt modern subjects who feel lost in the exchange between money and meaning, as shown by the works contributed by Freud (2003) on the uncanny, Bauman (2000) on liquid modernity, and Berger (Berger et al. 1974) on the homeless mind.

Unhomely Life generates two critical advancements to the geographical literature on home. First, scholars should not expect to find localized, coherent, and authentic construction of home culture in the context of speedy modernization. However, we should not treat alienation, uprootedness, and disjunction as the only parameters of modern times. No matter how much individuals feel alienated and uprooted, home remains a significant center of daily life through which people can live with a relentless world. Indeed, home becomes a meaningful site of imagination and experience to negotiate the interstice between constant mobility and much needed mooring. The making of home encompasses and embodies the multiple and contradictory spheres of life, torn apart by the tensions between profit and meaning, between control and resistance, and between here and there, so that people renegotiate their relationship with the broader society. The mundane experience of social differences in terms of ethnicity, age, class, and gender is intertwined with where and when to move, reflecting and reinforcing uneven mobilities. The differences all interlock, but individuals have variegated experiences of them in the process of traveling and mooring. A critical theory of home is rooted in the actual experiences of empowerment and disempowerment that seek to articulate and set free the nature of human beings, both internal and external, so that individuals can reshape the cultures of modern society for a balance between meaning and money.

Second, a temporal perspective is needed to unravel the projects of homemaking and unmaking. In contrast to a homogenous or transcendent temporality of the present, I suggest that the temporal perspective exemplifies a critical reading of continuity and disruption, highlighting what Walter Benjamin (1969, p. 262) calls the "constructive principle" of historiography, where "the historical materialist cannot do without the present which is not a transition, but in which time stands still and has to come to a stop. For this notion defines the present in which he himself is writing history." Here I elaborate two aspects to demonstrate how this perspective is valuable to the geography of home.

The journey-based approach aims to analyze an endless process of leaving one home, staying in another one at present, and longing for the next one in the future. From one home to another, the past does not disappear, the present is far from satisfying, and the future generates hope and uncertainty. Indeed, a journey to somewhere and somewhen becomes "an escape *for* home, not just *from* home" (Crouch 1994, p. 96, emphasis added), although these two homes entail different connotations. By inquiring into the life journeys of my respondents, I reveal a constant negotiation of the framing and naming of social reality, as people embrace either nostalgia or homesickness to express their dissonance with the way of living here and now. The past, as well as the home we leave behind, does not lie merely in memory, but continues to inform our current moment of transit in which continuity and disruption unfold. The literature on home thus needs to recognize that an endless life journey between here and there, and between past and present, constitutes an integral part of human experience.

Furthermore, the spiral dynamic between loafing and acceleration highlights the cultural value of home to enrich life journeys. In modern societies, clocks represent a powerful surveillance instrument to break up the natural cycle of time and orchestrate daily life from waking up to going to bed. This orchestration is so dominant that Lewis Mumford (2010, p. 14) emphasizes that the clock is essential to modern industrialization: "The clock, not the steam-engine, is the key-machine of the modern industrial age. For every phase of its development the clock is both the outstanding fact and the typical symbol of the machine: even today no other machine is so ubiquitous." Modernity in its spatiotemporal forms always carries its own paradoxical countermovement: closeness to natural environment, slowdown, disillusion with money economy, and so on. The surge of acceleration is followed by a countermovement of deceleration. The call for deceleration and the nostalgic desire for a slow world have never evaporated, but rather articulate alternative lifestyles, temporally embodied in slow movement, retreat, and loafing (Rosa 2013). Whether this call represents an overt resistance against the tyranny of time discipline is debatable; the return to and pursuit of a home always entails a powerful human agency, so that people avoid the fate of materialistic objectification and ultimate alienation. By bringing the use and meaning of time into the geographical scholarship on home, we can synthesize temporality, spatiality, and sociality in the representation and experience of home for a comprehensive understanding of home.

Remembering Home in China: By Whom and for What?

Modernization brought about the promise that China will become a strong and wealthy country in which individuals can enjoy a better life together and achieve the goal of common affluence, a goal rectified by the Chinese Communist Party's (CCP) top leaders as an essence of socialism. Following Deng Xiaoping's tutelage "time is money" and "efficiency is life," Chinese people have actively participated in the process of modernization with a hope of living their life with dignity and autonomy. For the first time, standardized clocks in factories and office towers replaced the natural cycles of rural seasonality in traditional China to orchestrate the daily life of hundreds of millions of people, compelling them to conform to time discipline and accelerate their activities around the dominant time schemes. The obsession with speed and growth that makes this process dominant, with prospects at once both empowering and disempowering, now renders daily life obsolete and meaningless and implicates individuals in a paradoxical condition between home and homelessness. In such a context, people seem to face difficulty in finding meaning in their daily life, and thus feel homeless in a relentless world. How can individuals enjoy a moderately prosperous standard of living (xiaokang shenghuo, 小康生活), as promised by the CCP under the leadership of President Xi Jinping, if they cannot call a place home? Once the promise of socialist modernization ends with meaninglessness, how can individuals counter their homeless mind?

The spiritual vacuum gives political elites to a reason to reorganize social and economic structures, such as a responsible society (Margaret Thatcher), the American dream (George W. Bush), or more recently, remembering hometown (Xi Jinping). In December 2013, one year after assuming the general secretary of the Central Committee of the CCP, President Xi Jinping made an important speech at a high-profile conference on urbanization in China. He emphasized that "it is necessary to rely on water and hills to integrate cities into natural environments so that residents can look upon hills, get close to water, and remember hometown." This is the first time that President Xi raised the idea of remembering hometown (jizhu xiangchou, 记住乡愁). In many instances that followed, he reiterated to the public the significance of remembering hometown. Rapidly, remembering hometown became a nationwide campaign to influence urban and rural development and draw much attention in social media. The examples range from visiting the hometown to look after elderly parents to environmental protection, from urban greening to traditional cultural rejuvenation. Under the guise of remembering hometown, numerous cultural and environmental projects have been launched in cities and villages to create spaces of remembering.

Semantically, *xiang* means hometown or soil, while *chou* refers to either feeling of homesickness or nostalgia. Although Fei Xiaotong's *xiangtu* (乡土) and Xi Jinping's *xiangchou* (乡愁) entail only one word difference, the difference exemplifies a great transformation in China. Spatially, the soil is replaced by high-rise buildings in cities as the image of Chinese society, and the soil is thus left behind as a forgotten land. Temporally, the natural cycle of time for agricultural planting and harvesting is replaced by time discipline and overtime work (such as 9-9-6 and 007, see Chapter 2) in factories and offices to orchestrate the everyday life of most of the Chinese population. Socially, soil-bound familiarity and peace of mind in the village give rise to strangeness and a homeless mind in the city. While displaced persons feel nostalgic for the good old days during their childhood in the countryside (see Chapter 5), people on the move constantly get stuck between homesickness for their places of origin and desire for an ideal home (see Chapters 3 and 4). What has happened in China is nothing new, akin to the great transformation in Europe after the Industrial Revolution. But the velocity and density of sociospatial transformation in post-Mao China is incomparable. For instance, the rate of urbanization jumped from 12.84% in 1953 and 17.58% in 1964 to 35.39% in 1990 and 49.68% in 2010 (National Bureau of Statistics 2021; Wu and Fang 2009). In 2020, the rate of permanent residents living in cities was 63.89%. In 1949, the number of residents in cities was 57.65 million, and then increased to 914.25 million in 2021 (National Bureau of Statistics 2022). The dizzying velocity of sociospatial transformation in China leaves many anxious about and frustrated at the relationship between the self and the world. Once cultural roots embedded in the soil are replaced by fleeting consumption fads, they feel lost in the high-speed train of modernization and cultivate a homeless mind.

The call of the soil, however, still lingers in China simply because Chinese society, as Fei Xiaotong emphasizes, is fundamentally rural. When rurality constitutes the foundations of Chinese society, and its glorious history of over 5,000 years has grown out of this soil, the call is natural and appealing. A sense of homeland (*xiangtu*, 乡土) has been rooted in China for centuries and become an essential part of Chinese culture. Pretty much like falling leaves returning to their roots, most Chinese people want their old bones to be buried in their hometown. The point is well taken. But given that the relentless process of modernization after 1978 displaces and dislocates individuals from their hometown into new and somehow hypermodern cities, how do we arrive at an age of permanent uprootedness? Between a soil-bound glorious history and an invincible trend of modernization, the whole country becomes unhomely on its own. That is why remembering hometown morphs into a nationwide campaign, in which the soil is not for permanent return, but for temporary remembering.

Underlining the campaign of remembering hometown is the risk that the hometown can be forgotten in the process of speedy modernization. Indeed, modernity and the accompanying expansion of the capitalist process of production and consumption generate cultural amnesia (Berman 1982). Forgetting is prioritized over remembering. Thus, Connerton (2009, p. 125) argues that forgetting is built into "the capitalist process of production itself, incorporated in the bodily experience of its life-spaces." In this case, the campaign of remembering hometown does not merely point to the need to retain the cultural meanings of home in a rapidly changing society, but also indicates a counter against forgetting, the culture logic of modernity, to promote public awareness of remembering the roots of Chinese culture. But the questions are: for whom, by what? Instead of placating individuals' restless soul, the political campaign of remembering hometown supports the extension of commercial exchange and political forces into the domestic spheres of everyday life.

For instance, remembering hometown means that individuals take care of their elderly parents and visit the village they leave behind once the hometown becomes the new frontier for social mooring to tie mobile individuals together. And if home is where parents live, anyone who actually wants to look after them should have adequate free time, which is virtually impossible in the race for wealth and fame. The Law of the People's Republic of China on Protection of the Rights and Interests of the Elderly (2012 Revision) stipulates the need for the sons and daughters of the elderly to go home frequently for caregiving. Article 18 specifies that "employers shall, in accordance with the relevant provisions of the state, ensure the rights of the supports to have the family visit leave."[2] The goal is to push younger generations to visit their elderly parents to fulfill the responsibility of filial piety and thus maintain traditional Chinese family values. The law has rarely been enacted to punish both employers and employees in defiance of the law, simply because time is not for home, but for money. Marx makes a great imaginative leap to grasp the desperate dynamism in the process of capital accumulation. As he puts it in the Communist Manifesto,

> Constant revolutionizing of production, uninterrupted disturbance of all social conditions, everlasting uncertainty and agitation distinguish the bourgeois epoch from all earlier ones. All fixed, fast frozen relations, with their train of ancient and venerable prejudices and opinions, are swept away, all new-formed ones become antiquated before they can ossify. All that is solid melts into air, all that is holy is profaned, and man is at last compelled to face with sober senses his real condition of life and his relations with his kind. (Marx and Engels 1988, p. 212)

Capital accumulation, for Marx, is indeed a universal destruction of the old social world. This exactly describes the current situation in post-Mao China. After the economic reform in 1978, the logic of capital accumulation ravages everyday life and plunders traditional community ties, forcefully dissolving the established social norms and cultural tradition to pave the way for the penetration of profit making into Chinese society.

Furthermore, the countryside, as the place of return and for remembering, has been hollowed out in the process of urbanization. Even those villages distant from Lijiang Old Town witness a dramatic demographic transformation, as younger generations move to the new city in Lijiang and make a living through the tourism industry rather than the soil. While increasing numbers of people choose to dwell and work in "modern" cities in China, rural villages and small towns are endowed with new meanings to attract urban residents to consume traditional idyllic lifestyles. For instance, Lijiang is highly regarded as a sensed home by migrants and tourists who have suffered from a hectic modern life. However, such a new home lacks the aura of familiarity that was integral to the old town in the past. Neither can it push individuals to root themselves in a same small, and even isolated, place (Fei 1992). In terms of the spatial and temporal qualities of rural life in China, rural villagers live their lives "in face-to-face intimacy and with repetitious and rigid life patterns" (Fei 1992, p. 59). At the dawn of modernization in the 1940s, Fei noted that these patterns made the rural village no longer a place where successful people in cities returned. Interestingly, eighty years later, Fei's observation is still accurate. Urban residents do not return to rural villages for long-term dwelling, but regard it as a consumption site for temporary relaxation and slowdown. The countryside, where today's middle-aged middle-class people lived during their childhood, is endowed with rural idyll in contrast with social acceleration and artificial concrete jungles in big Chinese cities. But this rooted home is no longer a place of permanent return, but only exists in their memories of childhood solidarity and closeness to a natural environment.

If attachment to the soil is a key feature of Chinese society and fosters China's cultural root, then industrialization and urbanization turn the soil into an unreturnable home, lost in speedy modernization. Thus, the campaign of "Remembering Hometown" is more like a nostalgic pursuit of what Fei calls familiarity and stability in the rural countryside, than a Maoist call for educated urban youth to receive reeducation and support rural development in villages in the 1960s and 1970s.[3] As Tinsley (2020, p. 2328) aptly describes, nostalgia reveals "a society's anxieties, its Others, and its temporal and spatial order of itself." Remembering hometown reflects such a nostalgic construction of the past in response to the loss of home in contemporary China. This sort of "imagined nostalgia," a representation through which people are motivated to yearn for a mediated world that they have never lost, becomes an effective way to project their roots back in both the past and the countryside, which elides many of the harsh features of the actual rural life that took place before and continues today. The accelerating circulation of capital, the expansion of industrialization and urbanization, and the spread of social anxiety and mental distress have facilitated the political institutionalization of nostalgia, as shown by the call for "remembering hometown."

In the past decade, the discourse of deceleration and slowdown has swelled as an individualistic reaction to hectic lifestyles in Chinese cities. Migration or traveling for a different lifestyle, particularly in Tibet and Yunnan, morphs into a strong crescendo among the middle-class in big Chinese cities. Books and movies devoted to a conscious slowdown in these places demonstrate such a popular appeal that a new consumption revolution emerges to reshape the relationship with food, sports, gardening, and so on. For instance, at JD.com, one of China's largest online shopping sites, a search for "slow life" results in over 6,000 books. Since 2011, "Slowness in the Past" (congqian man, 从前慢), a poem written by Mu Xin, has become popular among readers and been sung by some pop singers to provoke nationwide nostalgia for a lost tempo. While some people can manage their pace of daily life and control their own availability, some others engage in an endless rush to complete tasks or deliver packages. For instance, tens of thousands of delivery workers affiliated with JD.com are implicated in an algorithm intended to maximize efficiency: that is, delivering a book on slow life to customers as fast as possible. The acceleration of social life for efficiency and growth can thus make what was a sign of poverty and backwardness in rural villages, namely, slowness and following a natural cycle of time, an expensive luxury to be enjoyed by a small percentage of people in today's China.

The search for home, or remembering hometown, hides a dire fact: home culture is manufactured by those political elites who see the unruly reality and by those economic elites who discover its economic potential. The celebration of traditional lifestyle in China and the projection of home upon the countryside are synthesized into this campaign to entice individuals to believe that they can rest their soul elsewhere. Going home becomes a powerfully appealing melody to calm down these individuals in a relentless journey of making a living, but what is a place they can call home? When rooted home is unreturnable and sensed home is unreachable, a majority of ordinary people can become disillusioned with grand discourses, choose to lie down (tangping, 躺平), and withdraw from the race for wealth. Indeed, perhaps this is the right moment to return to Deng Xiaoping's decree of modernization, in which "getting rich first" means actually trading meaning for money, a deal of modernity that inspires individuals to become tourists, but turns them into unhomely subjects who struggle to reverse the deal for moments of sensed home.

Politically, the rhetoric of remembering home attempts to integrate a personal feeling of belonging into an imagined community of hundreds of millions, and in so doing creates a nationwide resonance between individual citizens and the entire nation. By portraying China as an ideal home and projecting this ideal onto a whole society, the rhetoric can justify the necessity of rational efficiency imposed on individuals in their daily life, not merely for their personal benefit, but also for a rising powerful country on the global stage. Nevertheless, this rhetoric resonates with personal desire to be rooted in a stable and coherent base so that individuals can feel at home, or at least remember that there is a place they can call home. Thus, home can be elevated from the micropolitics of everyday life to the macropolitics of nation building. In the wake of the 9/11 terrorist attacks in the United States, for instance, Walters (2004, p. 241) emphasizes that US citizens must protect their home as hearth or sanctuary against those who are uninvited, including illegal immigrants and bogus refugees, because "its contents (our property) are valuable and envied by others." Europe as a homespace becomes a political project for the European Union to justify border control and regulate immigrants, suggesting that the need for home is profoundly conservative, and even politically regressive (Mitzen 2018). Likewise, it is undoubted that the campaign of remembering hometown can generate profound implications for China's domestic politics and foreign relations. Home should be brought to the forefront in geographical analysis, and the making and unmaking of home constitutes a central topic in the conjunction between cultural politics and political economy.

Being Unhomely in Modern Times

> [The operations of] Heaven and Earth proceed in the most admirable way, but they say nothing about them; the four seasons observe the clearest laws, but they do not discuss them; all things have their complete and distinctive constitutions, but they say nothing about them. The sages trace out the admirable operations of Heaven and Earth, and reach to and understand the distinctive constitutions of all things; and thus it is that the Perfect Man (is said to) do nothing and the Greatest Sage to originate nothing, such language showing that they look to Heaven and Earth as their model. (Zhuangzi, Knowledge Rambling in the North, *Outer Chapters*)[4]

> Two things fill the mind with ever new and increasing admiration and awe, the more often and steadily we reflect upon them: the starry heavens above me and the moral law within me. I do not seek or conjecture either of them as if they were veiled obscurities or extravagances beyond the horizon of my vision; I see them before me and connect them immediately with the consciousness of my existence. (Kant 2010).

If people do not succumb to an imagined haven for eternality, but long for an actual world of cyclic and stable rhythms, we must listen to Zhuangzi, Kant, or other great scholars in the world. It is time to direct our attention to reconciling our relationship with the external nature (heaven and earth) and the internal nature (moral law and inner calm). These two excerpts, contributed by Zhuangzi and Kant, both point to the inherent connection between the two natures and tell us to place actual life within the conjunction so that we might not necessarily become eternal or perfect, but cultivate the consciousness of meaningful existence. As I have elaborated here, people project this consciousness onto the home, through the process of making and unmaking, to maintain the meanings of existence in a relentless world. Today, individuals rarely look to Heaven and Earth as their model of existence, but become uprooted from both natures as they must comply with socially constructed laws of time discipline and profit making. Uprootedness is one cause, if not the only cause, of the crisis of home in modern times, and explains the prevalence of unhomely life.

Sharma (2014) describes an encounter during her stayover in Atlanta's International Airport with a software developer for video games, who is a

Canadian citizen, but stationed in Buenos Aires. Echoing Bauman's idea of liquid life, the software developer defines himself as Liquid Man who relishes all the accouterments of a mobile, fluid, and busy lifestyle. To him, liquidity means no bonds in any specific place so that he can keep going with the flow without being limited by the weight of the world. Therefore, Sharma (2014, p. 29) laments that Liquid Man represents a subject who is "open, malleable, fragmented, unattached, and—of course—fast moving." While Sharma (2014) largely focuses on the uneven temporality constituted by Liquid Man and service workers in the airport, the implication for home is also explicit. It is not a house or apartment, but rather the airport, that concretizes Liquid Man's sense of home, a home where strangers accidentally gather for their next trips and where the constant is emptied out. To Liquid Man, home is not fixed, but fluid and experiential. Similarly, the cosmopolitan lady that Heller (1995) encountered in a jumbo jet already abandon the conception of home as solid or fixed. She can feel at home everywhere, but has no place to call home. What she experiences is a form of unhomely life. To Heller (1995, p. 2), this cosmopolitan lady is not a cultural monster against homo sapiens, but just a very lonely person, "one end product of two hundred years of modern history."

The modern history in the past two hundred years that Heller cites, describing the cosmopolitan lady, is largely a history of European industrialization and alienation. Due to the Enlightenment and mass industrialization, Europe entered the stage of modernization earlier than other continents and thus became exposed to the paradox of capitalist modernity. This paradox, according to Berman (1982), means that individuals were inspired by the promises of adventure, power, joy, growth, and great transformation, but realized that everything that they have and know could go in a different and even contradictory direction. For instance, Polanyi (2001) noticed a parallel existed between the ravages of the massive enclosures in English history and the social catastrophe deduced by the Industrial Revolution. As a matter of fact, economic improvements were "brought at the price of social dislocation" (Polanyi 2001, p. 79). To put it simply, the dominant trend in the process of modernization is that money has been prioritized over meaning. Once the over two hundred years of modernization are condensed into less than forty years in China, the great transformation captured by Polanyi is more dramatic and even traumatic. Regardless of the different velocities of transformation in Europe and China, one thing is certain. The pressure brought by speedy modernization causes endless anxiety, as individuals develop a mixed feeling of their hectic life. On the one hand, they are fearful of being left behind in the race for materialistic prosperity and want to make full use of time to gauge their economic return. On the other hand, they feel fatigued by the endless race and worried about whether they would be burnt out before they settle down. Their participation in and escape from the race creates a conundrum of in-between living between here and there, implicating them into an endless journey in search for home.

To avoid homelessness, individuals must foster and maintain a sense of home, sometimes by traveling elsewhere to experience a different time-space. Their longing for homely life coalesces into an escape from "the ambiguities and disappointments of everyday life" (Young 2005, p. 143). This longing is always for an elsewhere – rural countryside, exotic ethnic neighborhood, or cultural heritage sites – and an elsewhen – childhood, youth, or the ancients. Unhomely life descended on Europe in the past two hundred years, under the shadow of modernity, to activate an endless journey for home. As Heller observes in the context of Europe, one possibility is to dwell in a third home, or to visit it from time to time, which means that home is not fixed either here or there, but spatially and temporally located in between. The home–homelessness paring places individuals in an in-between condition of dwelling neither here nor there, yet also here and there, a condition reflecting Bhabha's (1994) critical term "unhomely life." When people live with an unhomely condition, they are dis-placed in various manners, but not re-placed in a fixed location, encountering a condition between here and there, familiarity and strangeness, through which they ambivalently experience a range of coexistences and nonexistences. Thus, being unhomely might enfold a permanent ambivalent conundrum of modernity, a conundrum leading beyond the duality of dwelling-traveling to think about the meanings of home in a highly mobile world.

Once we recognize that a pronouncement of homelessness generates a recurrent theme to highlight how the technological transformation of modern life, in line with the discipline of time and space, shapes social relations and mundane practices, we can explore the responses to such a loss by individuals in order to mediate speedy modernization. Even Derrida (2002) notes that modernity has provoked a desire for being at home. If there is a common theme to resonate with individuals from different backgrounds in the world, it can be "I want to be at home." Hence, Derrida (2002, p. 80) explains that "The more powerful and violent the technological expropriation, the delocalization, the more powerful, naturally, the recourse to the at-home, the return toward home." As Miller (2001, p. 4) contends, the home has become viewed as "a turbulent sea of constant negotiation rather than simply some haven for the self." For some, the turbulence comes from social discrimination against those in the marginal positions such as gays and lesbians; for some others, domestic violence remains a key source of turbulent oppression for women. Therefore, home becomes a site of and for ambiguity, simply because positive feelings of solidarity and safety provoked by home run parallel with severe acts of exclusion and oppression.

A critical analysis of home should involve connecting it to important issues at the core of political economy and cultural politics. While housing, property ownership, domestic chores, and family relations do matter, analyses must place home in the center of spatial and temporal organization in order to discuss the dynamic relationship between the self and the world, or how individuals can live a homely way of life in a relentless world. Beyond the material and domestic dimensions, home is thus endowed with theoretical significance to unravel those representations, experiences, and relations that reflect and reinforce the structural operations of modernity for the sake of capital accumulation and political control. The reality, however, is that homemaking becomes an unfulfilled project, in a state of constant crystallization and transgression, reflecting what Bauman (2000, p. 41) calls "vain and frustrating struggles to reforge their *de jure* status into the genuine autonomy and capacity for self-assertion." This is why I bring about the notion of unhomely life.

With respect to unhomely life, I seek to unpack an in-between condition that places individuals in a life journey to escape homelessness and search for home. Placing Zhuangzhi and Kant together, we might observe, interestingly, that in linking materialistic processes to the metaphysical question of being in the world, the unhomely unfolds in moments of existential anxiety. Simply put, both Zhuangzhi and Kant read heaven and the inner nature as the source of inspiration to launch a meaningful life journey from birth to death. Following their thoughts, we can ask a similarly existential question: How do we live a homely way of life to cope with the horror afflicted by relentless modernization after the Industrial Revolution? Or, how can we launch a journey to mitigate horror in our life? What is most discomforting, and sometimes inevitable, is the engulfment of this nature in the mechanical process of industrialization and modernization to such an extent that we are detached from our nature and thus do not feel at home in the world. This self-estranging experience of being not at home drives individuals to launch a return to a selective past, or travel to an idealized place for an experience of home, or live with homelessness. Regardless of the choice, the home remains an essential reference point to illuminate the meanings of human existence, not necessarily coming out of a determined fashion of profit and control, but rather an endless interaction between the self and the world.

Unhomely life remains a useful generic conception for understanding the complicated relationship between the self and the world. Here I reject the combination of housing and home that regards the pursuit of home as a struggle for stable housing, or as a resistance against domestic violence or household chores. I also reject uncritically taking for granted the loss of home as a particular feature of those who are houseless, as if the privileged class is bereft of such a loss – the supposed homely life constituted by one couple, two children, and one big house or apartment, giving rise to the fetishization of a certain family model. To understand and explore our life journey from there to here, from past to present, we need to look carefully into the subtle processes of homemaking and unmaking occurring at the planetary level, and unravel how unhomely life is a useful concept for understanding the in-between condition of living with speed and loafing, here and there, to reflect sociospatial inequality related to ethnicity, gender, age, and so on. By doing so, we can remain sensitive to the mundane (un) making of home in an era of speedy modernity.

Furthermore, the concept of unhomely life provides a framework for rethinking home and homelessness in ways that extend the discussion beyond established oppositions in the literature on home, including belonging versus alienation, here versus there, familiarity versus strangeness. The unhomely epitomizes a transit in which the framing and naming of social reality is negotiated: "not what lies inside or outside reality, but where to draw the 'meaningful' line between them" (Bhabha 1992, p. 144). Instead of thinking of being in the world as a static condition that neatly maps onto here or there, home or homeless, unhomely life makes it possible to consider an in-between being in terms of a plurality of representations and experiences within a rapidly changing society. The task is not to categorize home as a time-space of security and comfort distinct from that of homelessness as unsecured and unfamiliar, but to treat them as two sides of the coin of modernity and thus ask how individuals strive for living a homely life in a relentless world – representationally and experientially. It is unlikely, from this point of view, that it is possible to separate home from homelessness, simply because our life is a journey of stay and departure in which the home is a starting point and a destination of return. Unhomely life therefore brings to the fore the continual negotiation that lies at the heart of the politics of uneven mobilities.

Preparing for the coming and going of our life journeys, therefore, should not be such a big hassle in this highly mobile world. With electronic tickets ready in our smartphone, and suitcases well packed at the doorway, we should feel quite prepared to leave home and launch a new journey. The home becomes a point of departure, but where is the place we can call home? Or, do we have a destination to return to? The idea of home has always posed questions about who we are and where we rest our soul, and thus challenges the seemingly ordered world. This time, however, it is not *the* home, but homes, that push us to question ourselves: the place with comfort and pain, the movement from the past to the present and to the future, leaving home with "extreme feelings of abandonment and loss," as bell hooks (2009, p. 18) reflects on her unruly relationship with Kentucky, a place that she defines as native but dysfunctional. Due to this ambivalent nature of her homeplace, bell hooks (2009, p. 17) points out with pain that "My visits home almost always left me torn: I wanted to stay but I needed to leave, to be endlessly running away from home." Similarly, Bhabha (1994) wants to go home, to Bombay where he started his story, but admits that it is difficult for him to return to Mumbai. The mood of melancholy expressed by bell hooks and Bhabha is not uncommon, but has prevailed in modern times ever since the Industrial Revolution. The mood indicates that once the good (or bad) old days are gone forever, the destination for return is still there, but becomes unreturnable. While great scholars such as Zhuangzhi and Kant regard heaven and earth as the eternal reference point to lighten and guide a meaningful life journey, the reality is characterized by detachment from the soil and no time to contemplate the sky. Nevertheless, the call of home is still lingering and valuable. After all, in this relentless world, it becomes necessary to figure out a position in the swing between home and homelessness to cope with the deal of modernity.

Notes

1 http://english.www.gov.cn/news/top_news/2017/10/18/content_281475912475127.htm
2 http://www.lawinfochina.com/display.aspx?id=19790&lib=law
3 In 1968, Chairman Mao Zedong wrote, "It is very necessary for the educated youth to go to the countryside and undergo re-education by poor peasants." This instruction launched the Down to the Countryside Movement during the Cultural Revolution (1966–1976). Between 1968 and 1978, 17 million urban youth were estimated to migrate to remote villages and state-owned farms to work. For more details about the movement, see Zhou and Hou (1999).
4 Original text is天地有大美而不言，四时有明法而不议，万物有成理而不说。圣人者，原天地之美而达万物之理。是故至人无为，大圣不作，观于天地之谓也。Translated by James Legge. Source: https://ctext.org/zhuangzi/knowledge-rambling-in-the-north/ens

References

Adam, B. (1998). *Timescapes of Modernity: The Environment and Invisible Hazards*. London and New York: Routledge.

Adelman, H. (ed.) (2016). *Protracted Displacement in Asia: No Place to Call Home*. London: Routledge. https://doi.org/10.4324/9781315602752

Ahmed, S. (1999). Home and away: Narratives of migration and estrangement. *International Journal of Cultural Studies* 2 (3): 329–347. https://doi.org/10.1177/136787799900200303

Ahmed, S. (2000). *Strange Encounters: Embodied Others in Post-coloniality*. London and New York: Routledge.

Ahmed, S. (2004). Affective economies. *Social Text* 22 (2): 117–139.

Ahmed, S., Castaneda, C., Fortier, A.-M., and Sheller, M. (2003). *Uprootings/Regroundings: Questions of Home and Migration*. Oxford and New York: Berg.

Åkerlund, U. (2017). Strategic lifestyle management in later life: Swedish lifestyle movers in Malta seeking the "best of both worlds." *Population, Space and Place* 23 (1): e1964. https://doi.org/10.1002/psp.1964

Anderson, P. (1984). Modernity and revolution. *New Left Review March–April*: 96–113.

Angle, S.C. (2009). *Sagehood: The Contemporary Significance of Neo-Confucian Philosophy*. Oxford: Oxford University Press.

Appadurai, A. (1991). Global ethnoscapes: Notes and queries for a transnational anthropology. In: *Recapturing Anthropology: Working in the Present* (ed. R.G. Fox), 191–210. Santa Fe: School of American Research Press.

Appadurai, A. and Breckenridge, C.A. (1988). Why public culture? *Public Culture* 1 (1): 5–9. https://doi.org/10.1215/08992363-1-1-5

Arlt, W.G. (2013). The second wave of Chinese outbound tourism. *Tourism Planning & Development* 10 (2): 126–133. https://doi.org/10.1080/21568316.2013.800350

Unhomely Life: Modernity, Mobilities, and the Making of Home in China, First Edition. Xiaobo Su.

Ateljevic, I. and Doorne, S. (2000). "Staying within the fence": Lifestyle entrepreneurship in tourism. *Journal of Sustainable Tourism* 8 (5): 378–392. https://doi.org/10.1080/09669580008667374

Bachelard, G. (1964). *The Poetics of Space*. New York: Orion.

Bauman, Z. (1991). *Modernity and Ambivalence*. Cambridge: Polity.

Bauman, Z. (1997). *Postmodernity and its Discontents*. Cambridge: Polity.

Bauman, Z. (1998). *Globalization: The Human Consequences*. Cambridge: Polity.

Bauman, Z. (2000). *Liquid Modernity*. Cambridge: Polity.

Bauman, Z. (2004). *Wasted Lives: Modernity and its Outcasts (Reprint)*. Oxford: Polity.

Bauman, Z. (2005). Seeking shelter in Pandora's box. *City* 9 (2): 161–168. https://doi.org/10.1080/13604810500196949

Baxter, R. and Brickell, K. (2014). For home unmaking. *Home Cultures* 11 (2): 133–143. https://doi.org/10.2752/175174214X13891916944553

Baylina, M. and Gunnerud Berg, N. (2010). Selling the countryside: Representations of rurality in Norway and Spain. *European Urban and Regional Studies* 17 (3): 277–292. https://doi.org/10.1177/0969776409356215

Bell, D. and Hollows, J. (2007). Mobile homes. *Space and Culture* 10 (1): 22–39. https://doi.org/10.1177/1206331206296380

Bellezza, S., Paharia, N., and Keinan, A. (2017). Conspicuous consumption of time: When busyness and lack of leisure time become a status symbol. *Journal of Consumer Research* 44 (1): 118–138. https://doi.org/10.1093/jcr/ucw076

Benjamin, W. (1969). *Illumination*. New York: Shocken Books.

Bennett, K. (2011). Homeless at home in East Durham. *Antipode* 43 (4): 960–985. https://doi.org/10.1111/j.1467-8330.2010.00788.x

Benson, M. and O'Reilly, K. (2009). Migration and the search for a better way of life: A critical exploration of lifestyle migration. *The Sociological Review* 57 (4): 608–625. https://doi.org/10.1111/j.1467-954X.2009.01864.x

Benson, M. and O'Reilly, K. (2016). From lifestyle migration to lifestyle in migration: Categories, concepts and ways of thinking. *Migration Studies* 4 (1): 20–37. https://doi.org/10.1093/migration/mnv015

Benson, M. and Osbaldiston, N. (2014). *Understanding Lifestyle Migration: Theoretical Approaches to Migration and the Quest for a Better Way of Life*. London: Routledge.

Berger, P.L., Berger, B., and Kellner, H. (1974). *The Homeless Mind: Modernization and Consciousness*. New York: Vintage Books.

Bergmann, S. (2008). The beauty of speed or the discovery of slowness: Why do we need to rethink mobility? In: *The Ethics of Mobilities: Rethinking Place, Exclusion, Freedom and Environment* (ed. S. Bergmann and T. Sager), 13–24. Aldershot: Ashgate.

Berman, M. (1982). *All that is Solid Melts into Air: The Experience of Modernity (12th pr)*. London: Verso.

Besedovsky, N., Grafe, F.-J., Hilbrandt, H., and Langguth, H. (2019). Time as infrastructure: For an analysis of contemporary urbanization. *City* 23 (4–5): 580–588. https://doi.org/10.1080/13604813.2019.1689726

Betancur, J.J. (2002). The politics of gentrification: The case of West Town in Chicago. *Urban Affairs Review* 37 (6): 780–814. https://doi.org/10.1177/107874037006002

Bhabha, H. (1984). Of mimicry and man: The ambivalence of colonial discourse. *October* 28: 125–133. https://doi.org/10.2307/778467

Bhabha, H. (1992). The world and the home. *Social Text* 31/32: 141–153. https://doi.org/10.2307/466222

Bhabha, H. (1994). *The Location of Culture*. London and New York: Routledge.

Bianchi, R. (2006). Tourism and the globalisation of fear: Analysing the politics of risk and (in)security in global travel. *Tourism and Hospitality Research* 7 (1): 64–74. https://doi.org/10.1057/palgrave.thr.6050028

Blunt, A. (2007). Cultural geographies of migration: Mobility, transnationality and diaspora. *Progress in Human Geography* 31 (5): 684–694. https://doi.org/10.1177/0309132507078945

Blunt, A. and Bonnerjee, J. (2013). Home, city and diaspora: Anglo–Indian and Chinese attachments to Calcutta. *Global Networks* 13 (2): 220–240. https://doi.org/10.1111/glob.12006

Blunt, A. and Dowling, R.M. (2006). *Home*. New York: Routledge.

Blunt, A., Ebbensgaard, C.L., and Sheringham, O. (2021). The "living of time": Entangled temporalities of home and the city. *Transactions of the Institute of British Geographers* 46 (1): 149–162. https://doi.org/10.1111/tran.12405

Blunt, A. and Sheringham, O. (2019). Home-city geographies: Urban dwelling and mobility. *Progress in Human Geography* 43 (5): 815–834. https://doi.org/10.1177/0309132518786590

Blunt, A. and Varley, A. (2004). Geographies of home. *Cultural Geographies* 11 (1): 3–6. https://doi.org/10.1191/1474474004eu289xx

Boccagni, P. (2017). *Migration and the Search for Home: Mapping Domestic Space in Migrants' Everyday Lives*. New York: Palgrave Macmillan.

Bourdieu, P. (1984). *Distinction: A Social Critique of the Judgement of Taste (11th print)*. Cambridge, MA: Harvard University Press.

Brickell, K. (2012). "Mapping" and "doing" critical geographies of home. *Progress in Human Geography* 36 (2): 225–244. https://doi.org/10.1177/0309132511418708

Brickell, K. (2014). "Plates in a basket will rattle": Marital dissolution and home "unmaking" in contemporary Cambodia. *Geoforum* 51 (January): 262–272. https://doi.org/10.1016/j.geoforum.2012.12.005

Brickell, K. (2020). *Home SOS: Gender, Violence, and Survival in Crisis Ordinary Cambodia*. Hoboken, NJ: Wiley.

Brickell, K. and Datta, A. (2011). *Translocal Geographies: Spaces, Places, Connections*. London: Ashgate. https://www.taylorfrancis.com/books/e/9781315549910

Brown, M.F. (1996). On resisting resistance. *American Anthropologist* 98 (4): 729–735.

Buller, H. and Hoggart, K. (1994). *International Counterurbanization: British Migrants in Rural France*. Aldershot: Avebury.

Bureau of Statistics in Lijiang Old Town District. (2006). Statistic Bulletin of National Economy and Social Development in Lijiang Old Town District in 2005 (in Chinese). http://www.cnstats.org/tjgb/201003/ynljsgcq-2005-eoz_2.html (accessed 29 July 2021).

Bureau of Statistics in Lijiang Old Town District. (2021). Statistic Bulletin of National Economy and Social Development in Lijiang Old Town District in 2020 (in Chinese). http://ljgucheng.gov.cn/ljgc/c100025/202107/0cd57f37817c47bbb06d3c41f785449f/files/c71a0a53e1c54370aef1d77f1f1f92dc.pdf (accessed July 29 2021).

Butcher, J. (2003). *The Moralisation of Tourism: Sun, Sand … and Saving the World?* London and New York: Routledge.

Butcher, M. (2010). From "fish out of water" to "fitting in": The challenge of re-placing home in a mobile world. *Population, Space and Place* 16 (1): 23–36. https://doi.org/10.1002/psp.575

Cai, X. and Su, X. (2021). Dwelling-in-travelling: Western expats and the making of temporary home in Guangzhou, China. *Journal of Ethnic and Migration Studies* 47 (12): 2815–2832. https://doi.org/10.1080/1369183X.2020.1739392

Carson, D.A. and Carson, D.B. (2018). International lifestyle immigrants and their contributions to rural tourism innovation: Experiences from Sweden's far north. *Journal of Rural Studies* 64: 230–240. https://doi.org/10.1016/j.jrurstud.2017.08.004

Chambers, D. (2020). *Cultural Ideals of Home: The Social Dynamics of Domestic Space.* London: Routledge.

Chan, K.W. (2010). The household registration system and migrant labor in China: Notes on a debate. *Population and Development Review* 36 (2): 357–364. https://doi.org/10.1111/j.1728-4457.2010.00333.x

Chang, L.T. (2009). *Factory Girls: From Village to City in a Changing China.* New York: Spiegel & Grau.

Chang, W.-C. (2014). *Beyond Borders: Stories of Yunnanese Chinese Migrants of Burma.* Ithaca: Cornell University Press.

Chao, E. (2012). *Lijiang Stories: Shamans, Taxi Drivers, and Runaway Brides in Reform-era China.* Seattle: University of Washington Press.

Christensen, J. (2013). "Our home, our way of life": Spiritual homelessness and the socio-cultural dimensions of Indigenous homelessness in the Northwest Territories (NWT), *Canada. Social & Cultural Geography* 14 (7): 804–828. https://doi.org/10.1080/14649365.2013.822089

Chuluu, K.E. (2021). The Tongpo case: Indigenous institutions and environmental justice in China. *Critical Asian Studies* 53 (1): 109–125. https://doi.org/10.1080/14672715.2020.1854616

Clancy, M. (2017). *Slow Tourism, Food and Cities: Pace and the Search for the "Good Life."* New York: Routledge.

Clark, P. (1996). *The Chinese Cultural Revolution: A History.* Cambridge: Cambridge University Press.

Clifford, J. (1997). *Routes: Travel and Translation in the Late Twentieth Century.* Cambridge, MA: Harvard University Press.

Confucius (1967). *Book of Rites* (trans. J. Legge). New Hyde Park: University Books.

Connerton, P. (2009). *How Modernity Forgets.* Cambridge: Cambridge University Press.

Conrad, P. (1999). *Modern Times, Modern Places.* New York: Knopf.

Conradson, D. and Latham, A. (2005). Friendship, networks and transnationality in a world city: Antipodean transmigrants in London. *Journal of Ethnic and Migration Studies* 31 (2): 287–305. https://doi.org/10.1080/1369183042000339936

Crary, J. (2013). *24/7: Late Capitalism and the Ends of Sleep*. London: Verso.

Cresswell, T. (2006). *On the Move: Mobility in the Modern Western World.* New York: Routledge.

Crouch, D. (1994). Home, escape and identity: Rural cultures and sustainable tourism. *Journal of Sustainable Tourism* 2 (1–2): 93–101. https://doi.org/10.1080/09669589409510686

Cui, M. (2017). Reflections on Lijiang's heritage preservation (in Chinese). http://ynlj.wenming.cn/ljyaowen/201712/t20171219_4936988.html (accessed 29 September 2022).

Daniel, Y.P. (1996). Tourism dance performances authenticity and creativity. *Annals of Tourism Research* 23 (4): 780–797. https://doi.org/10.1016/0160-7383(96)00020-5

Darwall, S. (2006). The value of autonomy and autonomy of the will. *Ethics* 116 (2): 263–284. https://doi.org/10.1086/498461

de Certeau, M. (1984). *The Practice of Everyday Life*. Berkeley: University of California Press.

Deng, X. (1993). *Selections from Deng Xiaoping (Vol. III) (in Chinese)*. Beijing: Renming Chubanshe.

Derrida, J. (2002). *Echographies of Television: Filmed Interviews*. Cambridge: Polity Press.

Di Domenico, M. and Lynch, P.A. (2007). Host/guest encounters in the commercial home. *Leisure Studies* 26 (3): 321–338. https://doi.org/10.1080/02614360600898110

Douglas, M. (1991). The idea of a home: A kind of space. *Social Research* 58 (1): 287–307.

Drucker, P. (2001). The next society. *The Economist* (3–9 November): 3–20.

Duyvendak, J.W. (2011). *The Politics of Home: Belonging and Nostalgia in Western Europe and the United States*. Houndmills and New York: Palgrave Macmillan.

Easthope, H. (2004). A place called home. *Housing, Theory and Society* 21 (3): 128–138. https://doi.org/10.1080/14036090410021360

Edensor, T. (2012). The rhythms of tourism. In: *Real Tourism: Practice, Care, and Politics in Contemporary Travel Culture* (ed. C. Minca and T. Oakes), 68–85. Routledge.

Ehn, B. and Löfgren, O. (2010). *The Secret World of Doing Nothing*. Berkeley: University of California Press.

Escobar, A. (2001). Culture sits in places: Reflections on globalism and subaltern strategies of localization. *Political Geography* 20: 139–174.

Fanon, F. (1970). *Black SKIN, WHITE MASKS*. London: Paladin.

Fathi, M. (2021). "My life is on hold": Examining home, belonging and temporality among migrant men in Ireland. *Gender, Place & Culture*. https://doi.org/10.1080/0966369X.2021.1916445

Featherstone, M. (1987). Lifestyle and consumer culture. *Theory, Culture & Society* 4 (1): 55–70. https://doi.org/10.1177/026327687004001003

Ferguson, J. (1992). The Country and the City on the Copperbelt. *Cultural Anthropology*, 7 (1): 80–92.

Fei, H.-T. (1946). Peasantry and gentry: An interpretation of Chinese social structure and its changes. *American Journal of Sociology* 52 (1): 1–17.

Fei, X. (1992). *From the Soil: The Foundations of Chinese Society* (trans. G.G. Hamilton and Z. Wang). Berkeley: University of California Press.

Fludernik, M. and Nandi, M. (2014). *Idleness, Indolence and Leisure in English Literature.* Houndmills: Macmillan.

Fog Olwig, K. (2002). A wedding in the family: Home making in a global kin network. *Global Networks* 2 (3): 205–218. https://doi.org/10.1111/1471-0374.00037

Fog Olwig, K. and Hastrup, K. (eds.) (1997). *Siting Culture: The Shifting Anthropological Object.* London and New York: Routledge.

Foley, C. (2017). The art of wasting time: Sociability, friendship, community and holidays. *Leisure Studies* 36 (1): 1–20. https://doi.org/10.1080/02614367.2015.1055296

Foucault, M. (1979). *Discipline and Punish: The Birth of the Prison.* New York: Vintage Books.

Franklin, A. and Crang, M. (2001). The trouble with tourism and travel theory? *Tourist Studies* 1 (1): 5–22. https://doi.org/10.1177/146879760100100101

Freud, S. (1955). *The Standard Edition of the Complete Psychological Works of Sigmund Freud (Volume XVII).* London: The Hogarth Press.

Freud, S. (2003). *The Uncanny.* Penguin.

Friedman, R. (2020). *Modernity, Domesticity and Temporality in Russia: Time at Home.* London and New York: Bloomsbury Academic.

Fullagar, S., Markwell, K., and Wilson, E. (2012). *Slow Tourism: Experiences and Mobilities.* Bristol: Channel View Publications.

Gardiner, M. (1995). Utopia and everyday life in French social thought. *Utopian Studies* 6 (2): 90–123.

Geertz, C. (1963). *Agricultural Involution: The Processes of Ecological Change in Indonesia.* Berkeley: University of California Press.

Germann Molz, J. (2008). Global abode: Home and mobility in narratives of round-the-world travel. *Space and Culture* 11 (4): 325–342. https://doi.org/10.1177/1206331207308333

Giddens, A. (1990). *The Consequences of Modernity.* Cambridge: Polity Press.

Giddens, A. (1991). *Modernity and Self-identity: Self and Society in the Late Modern Age.* Stanford: Stanford University Press.

Gilman, C.P. (1903). *The Home: Its Work and Influence.* New York: The Co-Operative Press.

Ginsberg, R. (1999). Mediations on homelessness and being at home: In the form of a dialogue. In: *The Ethics of Homelessness* (ed. G.J.M. Abbarno), 29–40. Amsterdam: Rodopi.

Goullart, P. (1955). *Forgotten Kingdom.* London: John Murray.

Graham, S. (2004). Postmortem city. *City* 8 (2): 165–196. https://doi.org/10.1080/1360481042000242148

Gupta, A. and Ferguson, J. (1992). Beyond "culture": Space, identity, and the politics of difference. *Cultural Anthropology* 7 (1): 6–23. https://doi.org/10.1525/can.1992.7.1.02a00020

Haldrup, M. (2009). Banal tourism? Between cosmopolitanism and orientalism. In: *Cultures of Mass Tourism: Doing the Mediterranean in the Age of Banal Mobilities* (ed. P.O. Pons, M. Crang, and P. Travlou), 53–74. London: Routledge.

Hall, C.M., Lundmark, L., and Zhang, J.J. (eds.) (2021). *Degrowth and Tourism: New Perspectives on Tourism Entrepreneurship, Destinations and Policy*. London and New York: Routledge.

Hamilton, G.G. (1985). Why no capitalism in China? Negative questions in historical, comparative research. *Journal of Developing Societies* 1 (2): 187–211.

Hannam, K., Sheller, M., and Urry, J. (2006). Editorial: Mobilities, immobilities and moorings. *Mobilities* 1 (1): 1–22. https://doi.org/10.1080/17450100500489189

Hao, J., Zhu, J., and Thompson, S. (2022). Surviving in the post-repatriation era: Home-making strategies of homeless people in post-socialist China. *Housing Studies* 37 (2): 292–314. https://doi.org/10.1080/02673037.2020.1867082

Harari, Y.N. (2016). *Homo Deus: A Brief History of Tomorrow*. Toronto: Signal.

Hart, G.P. (2002). *Disabling Globalization: Places of Power in Post-apartheid South Africa*. Berkeley: University of California Press.

Harvey, D. (1982). *The Limits to Capital*. Oxford: Blackwell.

Harvey, D. (1989). *The Condition of Postmodernity: An Enquiry into the Origins of Cultural Change*. Oxford: Blackwell.

Harvey, D. (1996). *Justice, Nature, and the Geography of Difference*. Cambridge, MA: Blackwell Publishers.

Harvey, D. (2010a). *A Companion to Marx's Capital*. London and New York: Verso.

Harvey, D. (2010b). *The Enigma of Capital and the Crises of Capitalism*. Oxford and London: Oxford University Press.

He, S. and Wu, F. (2009). China's emerging neoliberal urbanism: Perspectives from urban redevelopment. *Antipode* 41 (2): 282–304. https://doi.org/10.1111/j.1467-8330.2009.00673.x

Heidegger, M. (1962). *Being and Time*. Oxford: Blackwell.

Heller, A. (1995). Where are we at home? *Thesis Eleven* 41 (1): 1–18. https://doi.org/10.1177/072551369504100102

Heller, A. (1999). *A Theory of Modernity*. Malden, MA: Blackwell.

Hemingway, J.L. (1996). Emancipating leisure: The recovery of freedom in leisure. *Journal of Leisure Research* 28 (1): 27–43. https://doi.org/10.1080/00222216.1996.11949759

Hinton, D. (ed.) (2005). *Mountain Home: The Wilderness Poetry of Ancient China*. New York: New Directions Book.

hooks, b. (2009). *Belonging: A Culture of Place*. New York: Routledge.

Hsee, C.K., Yang, A.X., and Wang, L. (2010). Idleness aversion and the need for justifiable busyness. *Psychological Science* 21 (7): 926–930. https://doi.org/10.1177/0956797610374738

Huang, L. and Xu, H. (2018). Therapeutic landscapes and longevity: Wellness tourism in Bama. *Social Science & Medicine* 197: 24–32. https://doi.org/10.1016/j.socscimed.2017.11.052

Huang, Y. (2008). *Capitalism with Chinese Characteristics: Entrepreneurship and the State*. Cambridge: Cambridge University Press.

Huang, Y. and Li, S. (2014). *Housing Inequality in Chinese Cities*. Hoboken: Taylor and Francis. http://public.ebookcentral.proquest.com/choice/publicfullrecord.aspx?p=1644446

Huang, Y., Yi, D., and Clark, W.A.V. (2020). Multiple home ownership in Chinese cities: An institutional and cultural perspective. *Cities* 97 (3). https://doi.org/10.1016/j.cities.2019.102518

Huq, E. and Harwood, S.A. (2019). Making homes unhomely: The politics of displacement in a gentrifying neighborhood in Chicago. *City & Community* 18 (2): 710–731. https://doi.org/10.1111/cico.12393

Jankowski, K.Z. (2018). The middling mobile: Finding place in the liquid city. *Mobilities* 13 (4): 601–614. https://doi.org/10.1080/17450101.2017.1403772

Juniu, S. (2000). Downshifting: Regaining the essence of leisure. *Journal of Leisure Research* 32 (1): 69–73. https://doi.org/10.1080/00222216.2000.11949888

Kant, I. (2010). *Critique of Practical Reason*. Amherst, NY: Prometheus Books.

Kern, L. (2016). Rhythms of gentrification: Eventfulness and slow violence in a happening neighbourhood. *Cultural Geographies* 23 (3): 441–457. https://doi.org/10.1177/1474474015591489

King, P. (2008). Memory and exile: Time and place in Tarkovsky's mirror. *Housing, Theory and Society* 25 (1): 66–78. https://doi.org/10.1080/14036090601151269

Kochan, D. (2016). Home is where I lay down my hat? The complexities and functions of home for internal migrants in contemporary China. *Geoforum* 71 (May): 21–32. https://doi.org/10.1016/j.geoforum.2016.02.021

Kraftl, P. (2007). Utopia, performativity, and the unhomely. *Environment and Planning D: Society and Space* 25 (1): 120–143. https://doi.org/10.1068/d397t

Kuhn, P.A. (2008). *Chinese among Others: Emigration in Modern Times*. Lanham, MD: Rowman & Littlefield.

Kundera, M. (1996). *Slowness*. New York: HarperCollins.

Kunz, S. (2016). Privileged mobilities: Locating the expatriate in migration scholarship. *Geography Compass* 10 (3): 89–101. https://doi.org/10.1111/gec3.12253

LaCouture, E. (2021). *Dwelling in the World: Family, House, and Home in Tianjin, China, 1860–1960*. New York: Columbia University Press.

Lam, T. (2022). The dark side of the miracle: Spectacular and precarious accumulation in an urban village under siege (A photo essay). *Positions* 30 (3): 523–547. https://doi.org/10.1215/10679847-9723711

Lasch, C. (1977). *Haven in a Heartless World: The Family Besieged*. New York: Basic Books.

Lefebvre, H. (1991a). *Critique of Everyday Life (Volume I)*. London and New York: Verso.

Lefebvre, H. (1991b). *The Production of Space*. Malden, MA: Blackwell.

Lefebvre, H. (2002). *Critique of Everyday Life (Volume II)*. London and New York: Verso.

Lefebvre, H. (2004). *Critique of Everyday Life (Volume III)*. London and New York: Verso.

Leonard, P. (2010). *Expatriate Identities in Postcolonial Organizations: Working Whiteness*. London: Routledge. https://www.taylorfrancis.com/books/e/9781315581453

Li, J. (2015). *Shanghai Homes: Palimpsests of Private Life*. New York: Columbia University Press.

Liang, S.Y. (2014). *Remaking China's Great Cities: Space and Culture in Urban Housing, Renewal, and Expansion*. Routledge. http://site.ebrary.com/id/10899387

Lin, G.C.S. (2014). China's landed urbanization: Neoliberalizing Politics, land commodification, and municipal finance in the growth of metropolises. *Environment and Planning A: Economy and Space* 46 (8): 1814–1835. https://doi.org/10.1068/a130016p

Lin, Y. (1998). *The Importance of Living*. New York: Harper.

Liu, C. (2021). Rethinking the timescape of home: Domestic practices in time and space. *Progress in Human Geography* 45 (2): 343–361. https://doi.org/10.1177/0309132520923138

Liu, Y., Wu, F., and He, S. (2008). The making of the new urban poor in transitional China: Market versus institutionally based exclusion. *Urban Geography* 29 (8): 811–834. https://doi.org/10.2747/0272-3638.29.8.811

Liu-Farrer, G. (2016). Migration as class-based consumption: The emigration of the rich in contemporary China. *The China Quarterly* 226: 499–518. https://doi.org/10.1017/S0305741016000333

Lu, S., Zhang, Y., and Wang, L. (2017). Social integration structure and determinant of tourism migration in ancient village: A case study of Hongcun village (in Chinese). *Tourism Tribune (Lvyou Xuekan)* 32 (4): 138–145.

Ma, H. and Liu, E. (2017). *Traditional Chinese Leisure Culture and Economic Development: A Conflict of Forces*. New York: Palgrave Macmillan.

Ma, S. and Xu, H. (2016). From consumption to production: Lifestyle migrants' living strategies in Dali old town (in Chinese). *Tourism Tribune (Lvyou Xuekan)* 31 (5): 81–88.

MacCannell, D. (1973). Staged authenticity: Arrangements of social space in tourist settings. *American Journal of Sociology* 79 (3): 589–603. https://doi.org/10.1086/225585

MacCannell, D. (1976). *The Tourist: A New Theory of the Leisure Class*. Berkeley: University of California Press.

Mallett, S. (2004). Understanding home: A critical review of the literature. *The Sociological Review* 52 (1): 62–89. https://doi.org/10.1111/j.1467-954X.2004.00442.x

Marcuse, H. (1964). *One-Dimensional Man*. Boston: Beacon Press.

Marks, L. (2002). Maldives: Time stands still. *The Telegraph* 30 April. http://www.telegraph.co.uk/travel/destinations/africaandindianocean/maldives/724122/Maldives-Time-stands-still.html

Marx, K. and Engels, F. (1988). *The Communist Manifesto*. Amherst, NY: Prometheus Books.

Mason, K.A. (2020). Chrono-socialism with Chinese characteristics: Temporal balancing and moralities of time in contemporary urban China. *American Anthropologist* 122 (4): 852–863. https://doi.org/10.1111/aman.13475

Massey, D. (1994). *Space, Place and Gender*. Minneapolis: University of Minnesota Press. https://nbn-resolving.org/urn:nbn:de:101:1-201410235969

Massey, D. (2007). *World City*. Cambridge: Polity Press.

Mauss, M. (1990). *The Gift: The Form and Reason for Exchange in Archaic Societies*. London: Routledge.

May, J. and Thrift, N. (2001). *Timespace: Geographies of Temporality*. London: Routledge.

May, T. (1994). The concept of autonomy. *American Philosophical Quarterly* 31 (2): 133–144.

McCarthy, L. (2018). (Re)conceptualising the boundaries between home and homelessness: The *unheimlich*. *Housing Studies* 33 (6): 960–985. https://doi.org/10.1080/02673037.2017.1408780

McClintock, A. (1995). *Imperial Leather: Race, Gender, and Sexuality in the Colonial Contest*. London: Routledge.

McDowell, L. (1999). *Gender, Identity, and Place: Understanding Feminist Geographies*. Minneapolis: University of Minnesota Press.

McQuire, S. (2008). *The Media City: Media, Architecture and Urban Space*. London: Sage.

Miller, D. (2001). *Home Possessions: Material Culture Behind Closed Doors*. New York: Berg.

Mills, C.W. (1970). *The Sociological Imagination*. Harmondsworth: Penguin.

Mitzen, J. (2018). Feeling at home in Europe: Migration, Ontological security, and the political psychology of EU bordering. *Political Psychology* 39 (6): 1373–1387. https://doi.org/10.1111/pops.12553

Morley, D. (2000). *Home Territories: Media, Mobility and Identity*. London: Routledge.

Morley, D. and Robins, K. (1995). *Spaces of Identity: Global Media, Electronic Landscapes and Cultural Boundaries*. London: Routledge.

Mumford, L. (2010). *Technics and Civilization*. Chicago: University of Chicago Press.

National Bureau of Statistics (2021). *China Statistical Yearbook 2021*. Beijing: China Statistics Press.

National Bureau of Statistics (2022). Statistic bulletin of China's national economy and social development in 2021 (in Chinese). http://www.stats.gov.cn/tjsj/zxfb/202202/t20220227_1827960.html (accessed 13 July 2022).

Nguyen, V. (2017). Slow construction. *City* 21 (5): 650–662. https://doi.org/10.1080/13604813.2017.1375728

Nowicka, M. (2007). Mobile locations: Construction of home in a group of mobile transnational professionals. *Global Networks* 7 (1): 69–86. https://doi.org/10.1111/j.1471-0374.2006.00157.x

Oakes, T. (1998). *Tourism and Modernity in China*. London: Routledge.

O'Reilly, K. and Benson, M. (2009). Lifestyle migration: Escaping to the good life? In: *Lifestyle Migration: Expectations, Aspirations and Experiences* (ed. M. Benson and K. O'Reilly), 1–13. Burlington, VT: Ashgate.

Osborne, P. (2000). *Traveling Light: Photography, Travel, and Visual Culture*. Manchester: Manchester University Press.

Oshana, M. (2006). *Personal Autonomy in Society*. Aldershot and Burlington, VT: Ashgate.

Peters, H.A. (2013). Dancing in the market: Reconfiguring commerce and heritage in Lijiang. In: *Cultural Heritage Politics in China* (ed. T. Blumenfield and H. Silverman), 115–140. New York: Springer. https://doi.org/10.1007/978-1-4614-6874-5_7

Pine, B.J., and Gilmore, J.H. (1998). Welcome to the experience economy. *Harvard Business Review* 1998 (July–August): 97–105.

Polanyi, K. (2001). *The Great Transformation*. Boston: Beacon Press.

Pratt, G. (2004). *Working Feminism*. Philadelphia: Temple University Press.

Pratt, M.L. (1992). *Imperial Eyes: Travel Writing and Transculturation*. London: Routledge.

Pun, N. (2005). *Made in China: Women Factory Workers in a Global Workplace*. Durham, NC: Duke University Press.

Qian, Z. (1997). *Qian Zhongshu san wen (Qian Zhongshu's Selected Essays)*. Hangzhou: Zhejiang wenyi chubanshe.

Rapport, N. and Dawson, A. (eds.) (1998). *Migrants of Identity: Perceptions of Home in a World of Movement*. Oxford: Berg.

Rees, H. (2000). *Echoes of History: Naxi Music in Modern China*. Cary: Oxford University Press.

Rojek, C. (1997). Indexing, dragging and the social construction of tourist sites. In: *Touring Cultures: Transformations of Travel and Theory* (ed. C. Rojek and J. Urry), 52–74. London and New York: Routledge.

Rosa, H. (2013). *Social Acceleration: A New Theory of Modernity*. New York: Columbia University Press.

Rosa, H., Dörre, K., and Lessenich, S. (2017). Appropriation, activation and acceleration: The escalatory logics of capitalist modernity and the crises of dynamic stabilization. *Theory, Culture & Society* 34 (1): 53–73. https://doi.org/10.1177/0263276416657600

Rosa, H. and Scheuerman, W. (2010). *High-speed Society: Social Acceleration, Power, and Modernity*. University Park, PA: Penn State Press.

Rozelle, S. and Hell, N. (2020). *Invisible China: How the Urban-Rural Divide Threatens China's Rise*. Chicago: University of Chicago Press.

Sanders, S.R. (1993). *Staying Put: Making a Home in a Restless World*. Boston: Beacon Press.

Saunders, P., and Williams, P. (1988). The constitution of the home: Towards a research agenda. *Housing Studies* 3 (2): 81–93. https://doi.org/10.1080/02673038808720618

Schlecker, M. (2005). Going back a long way: "Home place," thrift and temporal orientations in northern Vietnam. *Journal of the Royal Anthropological Institute* 11 (3): 509–526. https://doi.org/10.1111/j.1467-9655.2005.00248.x

Seaton, J.P. and Maloney, D. (1994). *A Drifting Boat: An Anthology of Chinese Zen Poetry*. Buffalo, NY: White Pine Press.

Sharma, S. (2014). *In the Meantime: Temporality and Cultural Politics*. Durham: Duke University Press.

Sheffer, G. (2003). *Diaspora Politics: At Home Abroad*. Cambridge: Cambridge University Press.

Shell, S.M. (2009). *Kant and the Limits of Autonomy*. Cambridge, MA: Harvard University Press.

Sheller, M. (2016). Uneven mobility futures: A Foucauldian approach. *Mobilities* 11 (1): 15–31. https://doi.org/10.1080/17450101.2015.1097038

Simmel, G. (1997). *Simmel on Culture: Selected Writings*. London: Sage.

Skeggs, B. (2004). *Class, Self, Culture*. London: Routledge.

Smith, M K. and Kelly, C. (2006). Holistic tourism: Journeys of the self? *Tourism Recreation Research* 31 (1): 15–24. https://doi.org/10.1080/02508281.2006.11081243

Spangler, I. (2020). Hidden value in the platform's platform: Airbnb, displacement, and the un-homing spatialities of emotional labour. *Transactions of the Institute of British Geographers* 45: 575–588. https://doi.org/10.1111/tran.12367

Su, X. (2011). Heritage production and urban locational policy in Lijiang, China. *International Journal of Urban and Regional Research* 35 (6): 1118–1132. https://doi.org/10.1111/j.1468-2427.2010.01011.x

Su, X. (2014). Tourism, modernity and the consumption of home in China. *Transactions of the Institute of British Geographers* 39 (1): 50–61. https://doi.org/10.1111/tran.12013

Su, X. and Chen, Z. (2017). Embeddedness and migrant tourism entrepreneurs: A Polanyian perspective. *Environment and Planning A: Economy and Space* 49 (3): 652–669. https://doi.org/10.1177/0308518X16674337

Su, X. and Teo, P. (2009). *The Politics of Heritage Tourism in China: A View from Lijiang*. London: Routledge. https://doi.org/10.4324/9780203873687

Sun, X. and Xu, H. (2017). Lifestyle tourism entrepreneurs' mobility motivations: A case study on Dali and Lijiang, China. *Tourism Management Perspectives* 24: 64–71. https://doi.org/10.1016/j.tmp.2017.06.004

Sullivan, O. (2008). Busyness, status distinction and consumption strategies of the income rich, time poor. *Time & Society* 17 (1): 5–26. https://doi.org/10.1177/0961463X07086307

Swartz, W. (2008). *Reading Tao Yuanming: Shifting Paradigms of Historical Reception (427–1900)*. Cambridge, MA: Harvard University Press.

Tang, S., Zhou, J., Lin, S., and Li, X. (2022). Where is my home? Sense of home among rural migrant women in contemporary China. *Geoforum* 129: 131–140. https://doi.org/10.1016/j.geoforum.2022.01.014

Thomas, L. (2008). Alternative realities: Downshifting narratives in contemporary lifestyle television. *Cultural Studies* 22 (5): 680–699. https://doi.org/10.1080/09502380802245936

Thompson, E.P. (1967). Time, work-discipline and industrial capitalism. *Past and Present* 38 (1): 56–97.

Tian, X. (2005). *Tao Yuanming & Manuscript Culture: The Record of a Dusty Table*. Seattle: University of Washington Press.

Tinsley, M. (2020). Revisiting nostalgia: Imperialism, anticolonialism, and imagining home. *Ethnic and Racial Studies* 43 (13): 2327–2355. https://doi.org/10.1080/01419870.2020.1727935

Tomlinson, J. (2007). *The Culture of Speed*. London: Sage.

Torres, F.T. (2021). *Temporal Regimes: Materiality, Politics, Technology*. London: Routledge.

Tsai, K.S. (2000). Banquet banking: Gender and rotating savings and credit associations in South China. *The China Quarterly* (161): 142–170.

Tu, W. (2001). The ecological turn in new Confucian humanism: Implications for China and the world. *Daedalus* 130 (4): 243–264.

Tuan, Y.-F. (1977). *Space and Place: The Perspective of Experience*. Minneapolis: University of Minnesota Press.

Tuan, Y.-F. (1980). Rootedness versus sense of place. *Landscape* 24: 3–8.

Tuan, Y.-F. (1990). *Topophilia: A Study of Environmental Perception, Attitudes, and Values*. New York: Columbia University Press.

Tuan, Y.-F. (2007). *Coming Home to China*. Minneapolis: University of Minnesota Press.

Tucker, A. (1994). In search of home. *Journal of Applied Philosophy* 11 (2): 181–187. https://doi.org/10.1111/j.1468-5930.1994.tb00107.x

Turner, V. (1973). The center out there: Pilgrim's goal. *History of Religions* 12 (3): 191–230. https://doi.org/10.1086/462677

Tynen, S. (2019). Lived space of urban development: The everyday politics of spatial production in Nanjing, *China. Space and Culture* 22 (2): 172–188. https://doi org/10.1177/1206331218774480

Urry, J. (2002). *The Tourist Gaze* (2nd ed.). London: Sage.

van Liempt, I. and Staring, R. (2021). Homemaking and places of restoration: Belonging within and beyond places assigned to Syrian refugees in the Netherlands. *Geographical Review* 111 (2): 308–326. https://doi.org/10.1080/00167428.2020.1827935

Veblen, T. (2007[1899]). *The Theory of the Leisure Class*. Oxford: Oxford University Press.

Vidler, A. (1992). *The Architectural Uncanny: Essays in the Modern Unhomely (5th print)*. Cambridge, MA: MIT Press.

Virilio, P. (1986). *Speed and Politics: An Essay on Dromology*. New York: Semiotext.

Wajcman, J. (2015). *Pressed for Time: The Acceleration of Life in Digital Capitalism*. Chicago: University of Chicago Press.

Walters, W. (2004). Secure borders, safe haven, domopolitics. *Citizenship Studies* 8 (3): 237–260. https://doi.org/10.1080/1362102042000256989

Wang, H. (2009). *The End of the Revolution: China and the Limits of Modernity*. London: Verso.

Wang, K. and He, S. (2019). Unraveling the marginalization of new generation peasant workers in China: Cultural reproduction and symbolic construction. *Journal of Urban Affairs* 41 (3): 282–304. https://doi.org/10.1080/07352166.2018.1451253

Wang, M. (2018). "Rigid demand": Economic imagination and practice in China's urban housing market. *Urban Studies* 55 (7): 1579–1594. https://doi.org/10.1177/0042098017747511

Wang, M. (2022). Calculating needs and prescribing properties: Chaiqian and the commensuration of value at Nanjing's urban edge. *Positions* 30 (3): 501–521. https://doi.org/10.1215/10679847-9723698

Wang, N. (1999). *Tourism and Modernity: A Sociological Analysis*. New York: Pergamon.

Wang, X., Bai, K., and Bao, J. (2021). Dimension construction and test of place integration of tourism migration in Lijiang ancient town (in Chinese). *Tourism Tribune (Lvyou Xuekan)* 36 (6): 116–132.

Wang, Y.P. (2004). *Urban Poverty, Housing and Social Change in China*. London and New York: Routledge.

Wang, Y.P., and Murie, A. (2000). Social and Spatial Implications of Housing Reform in China. *International Journal of Urban and Regional Research* 24 (2): 397–417. https://doi.org/10.1111/1468-2427.00254

Wardhaugh, J. (1999). The unaccommodated woman: Home, homelessness and identity. *The Sociological Review* 47 (1): 91–109. https://doi.org/10.1111/1467-954X.00164

Wei, S.-J., Zhang, X., and Liu, Y. (2017). Home ownership as status competition: Some theory and evidence. *Journal of Development Economics* 127: 169–186. https://doi.org/10.1016/j.jdeveco.2016.12.001

Werner, S.E. (2012). *After work or study abroad: Chinese return migration and Kunming's "Jia Xiang Bao": hometown babies (Doctoral dissertation)*. University of Minnesota, Minneapolis.

Wilcox, K., Laran, J., Stephen, A.T., and Zubcsek, P.P. (2016). How being busy can increase motivation and reduce task completion time. *Journal of Personality and Social Psychology* 110 (3): 371–384. https://doi.org/10.1037/pspa0000045

Wiles, J. (2008). Sense of home in a transnational social space: New Zealanders in London. *Global Networks* 8 (1): 116–137. https://doi.org/10.1111/j.1471-0374.2008.00188.x

Wong, S.W., Tang, B., and Liu, J. (2022). Rethinking China's rural revitalization from a historical perspective. *Journal of Urban History* 48 (3): 565–577. https://doi.org/10.1177/0096144220952091

Wright, T. (2010). *Accepting Authoritarianism: State–Society Relations in China's Reform Era*. Stanford: Stanford University Press. https://doi.org/10.1515/9780804774253

Wu, F. (2018). Planning centrality, market instruments: Governing Chinese urban transformation under state entrepreneurialism. *Urban Studies* 55 (7): 1383–1399. https://doi.org/10.1177/0042098017721828

Wu, F., Zhang, F., and Webster, C.J. (2014). *Rural Migrants in Urban China: Enclaves and Transient Urbanism*. New York: Routledge.

Wu, K. and Fang, C. (2009). The development process and basic pattern of China's small towns since 1949 and its recent new situation (in Chinese). *Economic Geography (Jingji Dili)* 29 (10): 1605–1611.

Wu, Y. and Yang, J. (2022). Lifestyle skiing contributes to health and eudaimonic well-being: An emergent Chinese lifestyle mobility. *Tourism Geographies* 24 (6–7): 1134–1152. https://doi.org/10.1080/14616688.2021.1955958

Xu, H., Ma, S., and Jiang, L. (2017). Social interactions of lifestyle tourism entrepreneurial migrants (in Chinese). *Tourism Tribune (Lvyou Xuekan)* 32 (7): 69–76.

Xu, H. and Wu, Y. (2016). Lifestyle mobility in China: Context, perspective and prospects. *Mobilities* 11 (4): 509–520. https://doi.org/10.1080/17450101.2016.1221027

Yan, Y. (2009). *The Individualization of Chinese Society*. Oxford: Berg.

Yang, A.X. and Hsee, C.K. (2019). Idleness versus busyness. *Current Opinion in Psychology* 26: 15–18. https://doi.org/10.1016/j.copsyc.2018.04.015

Yang, M.M. (2004). Spatial struggles: Postcolonial complex, state disenchantment, and popular reappropriation of space in rural southeast China. *The Journal of Asian Studies* 63 (3): 719–755. https://doi.org/10.1017/S002191180400169X

Young, I.M. (1997). *Intersecting Voices*. Princeton: Princeton University Press.

Young, I.M. (2005). *On Female Body Experience: "Throwing Like a Girl" and Other Essays*. Oxford: Oxford University Press.

Yunnan Committee for Five Series of Books on Ethnic Issues (1983). *Socio-historical Investigation on Naxi Ethnicity (Naxizu shehui lishi diaocha)*. Kunming: Yunnan minzhu chubanshe.

Zhan, Y. (2022). "When it's dark in the east, it's light in the west": Lifelong venturing and accelerated temporality in Beijing's urban villages. *Positions* 30 (3): 595–617. https://doi.org/10.1215/10679847-9723750

Zhang, J. (2021). Utopian myth among lifestyle migrant businesspersons (in Chinese). *Tourism Tribune (Lvyou Xuekan)* 36 (10): 7–9. https://doi.org/10.19765/j.cnki.1002-5006.2021.10.004

Zhang, L. (2012). *In Search of Paradise: Middle-class Living in a Chinese Metropolis*. Ithaca, NY: Cornell University Press.

Zhang, L. (2018). Cultivating the therapeutic self in China. *Medical Anthropology* 37 (1): 45–58. https://doi.org/10.1080/01459740.2017.1317769

Zhang, L. (2020). *Anxious China: Inner Revolution and Politics of Psychotherapy*. Oakland, CA: University of California Press.

Zhao, X. (2013). The development and change of the Cong among the Naxi in Southwest China (in Chinese). *Journal of Ethnology* 4 (5): 25–33.

Zhou, X. and Hou, L. (1999). Children of the Cultural Revolution: The state and the life course in the People's Republic of China. *American Sociological Review* 64 (1): 12–36. https://doi.org/10.2307/2657275

Zhu, Y. (2018). *Heritage and Romantic Consumption in China*. Leiden: Amsterdam University Press.

Zukin, S. (2009). *Naked City: The Death and Life of Authentic Urban Places*. Oxford: Oxford University Press.

Index

Unhomely Life: Modernity, Mobilities, and the Making of Home in China, First Edition. Xiaobo Su.